Electromagnetic Behaviour of Metallic Wire Structures

S. T. Chui · Lei Zhou

Electromagnetic Behaviour of Metallic Wire Structures

 Springer

S. T. Chui
Department of Physics and Astronomy
University of Delaware
Newark, NJ
USA

Lei Zhou
Department of Physics
Fudan University
Shanghai
China

ISBN 978-1-4471-5802-8   ISBN 978-1-4471-4159-4 (eBook)
DOI 10.1007/978-1-4471-4159-4
Springer London Heidelberg New York Dordrecht

Printed on acid-free paper

Springer is part of Springer Science+Business Media (www.springer.com)

Preface

Over the last 10 years there has been a lot of exciting development in the photonics of metallic wire structures. The concept and the physics involved are quite simple. Yet this message is usually obscured and not explained and emphasized. In this book we try to present our point of view on how these phenomena can be understood in a simple way without carrying out detailed numerical calculations every time. It turns out that the method presented here also provides for a very efficient algorithm for dealing with electromagnetic waves in these structures. We have tried to make the book simple enough so that it will be useful for non-specialists in other fields to enter into this area. This book is intended for a general audience in both physics and engineering. To include the possibility of using this as a textbook, we have provided detailed derivations of different formulae. We have tried to make the book self-contained by providing enough background information so that it is not necessary to chase through different sources to understand and follow the development of the ideas and the derivation of the formulae.

STC would like to thank his colleagues for helpful discussion and who provided key ideas presented in this book. In particular, he would like to thank Prof. Zhifang Lin, Prof. John Xiao, and Prof. Weiyi Zhang. He thanks the US DOE and NASA for financial support. ZL thanks Mr. Che Qu for help in numerical simulations and the NSFC for financial support.

Contents

Chapter 1
Electromagnetic Waves in Metallic Wire Structures

1.1 Introduction

With the advance of nanofabrication technology, more and more artificial materials have been proposed and developed in the past 20 years. Two representative examples are photonic and phononic crystals [1–4]. The existence of the forbidden gap makes it possible to manipulate and control the flow of electromagnetic (EM) and acoustic waves. More recently, the focus has switched to metallic wire structures. Macroscopic metallic wire structures have been much studied in the past, partly connected with antenna applications. Recently, EM composite materials with metallic wire components have received much attention. Metallic wire structures such as fractals have been shown to be efficient wide band filters and ultra-compact reflectors [5–7]. Man-made metamaterials with artificial magnetism [8, 9] and negative refraction index [10, 11] were designed and experimentally verified. The original negative-index material [10, 11] consists of straight metallic wires and metallic split-ring resonators (SRRs), and the purpose is to adjust the electric dipole eigenmode in the wire and the "magnetic" eigenmode in the SRR, so that their out-of-phase responses have an overlapping frequency region. Since then, near-field subwavelength images have been realized by several groups with different types of metamaterials [12–15], based on early theoretical proposals [16, 17]. The possibility of making metallic wire structures of sizes less than microns opens the door to explore new physical phenomena. Micron size helixes may exhibit a giant (orders of magnitude larger than current materials) Faraday rotation at infrared frequencies [18]. The possibility of making helixes at micron scales enables the recent experimental study of these as efficient circular polarizers at the far infrared [19]. Some metallic wire structures possess rather intriguing EM properties. While the magnetization (electric polarization) in ordinary materials is generally induced by an external magnetic (electric) field, in some particular metallic wire structures, an external electric (magnetic) field can induce a magnetic (electric) polarization. Materials that are both ferromagnetic and ferroelectric at the same time (multiferroics) exhibit this type of phenomena which

S. T. Chui and L. Zhou, *Electromagnetic Behaviour of Metallic Wire Structures*,
DOI: 10.1007/978-1-4471-4159-4_1, © Springer-Verlag London 2013

is called magnetoelectric effect. Interests in this type of materials have recently revived due to improved sophistication in creating multiphase nanostructure material. While the magnetoelectric effect in multiferroic materials decreases drastically above the spin wave frequency, if the wire structures are small enough the magnetoelectric effect can persists up to much higher frequencies. These magnetoelectric materials can exhibit novel EM properties. For example, the Poynting vector can be perpendicular to the wave vector, and the EM eigenmodes inside such a medium can be quire nontrivial [20].

The fascinating phenomena mentioned above occur in composite materials involving metallic wire structures. Knowledge of the EM eigenmodes in metallic wire structures is crucial to understand these physical effects. Full-wave numerical simulations based on finite-difference time-domain (FDTD) and finite-element method were frequently adopted to reproduce the observed phenomena. Such calculations shed very little light on the inherent physics behind the physical phenomena. We have recently developed a rigorous and general circuit theory that provides a modular, simple, and physical way to understand wire structures with metallic components. Furthermore, we find that quite often this approach is numerically much more efficient than current algorithms. In this book, we explain this theoretical framework to study the EM eigenmodes in arbitrary metallic wire structures, and apply the theory to various examples to illustrate our understandings of the inherent physics behind the physical phenomena occurring in such medium.

A circuit theory includes circuit parameters such as the inductances and the capacitances of the circuit. In general, the current distribution along the wires is not uniform and the circuit elements cannot be approximated by a *single* "local" (self) lumped elements; the modulation along the current path has to be taken into consideration. Two different routes have been explored [21–23, 18, 24, 25]. One is the "partial equivalent element circuit" (PEEC) method that discretizes the inductances and capacitances into small elements distributed in space [21] and considers the self and mutual circuit elements between them. We have recently developed another method by expanding the current distribution along the wire in terms of a Fourier series [18, 22–25]. We found that this latter approach provides for a simpler, more physical, and numerically very efficient way to understand metallic wire networks. In some cases, the problem becomes analytically tractable. For example, there is a folklore that, for a metallic wire of arbitrary shape of length L, the resonance wavelengths λ_r is close to twice its length divided by an integer, $\lambda_r = 2L/n$, but no proof has ever been given. In our approach, we can show that this is true in the thin-wire limit. Ways to calculate the corrections as the wire gets thicker are demonstrated. We summarize and explain briefly here the advantages of the nonlocal basis functions that we adopted. The details will be presented in different chapters following this introduction.

(1) For thin wires, the "self" circuit impedances (inductances and capacitances) between the same Fourier modes are of the order of *log (radius of wire/length of wire)* and much larger than that ("mutual") between different Fourier

modes, which are *not* log divergent. The circuit problem generally involves inverting the circuit impedance matrix to relate the induced current in the wires to the externally applied electric field. In the Fourier basis, the matrix is approximately diagonal; its inversion is easier and the results are easier to understand. The circuit equations are simplified greatly.

(2) In this basis, the value of the impedance matrix element increases rapidly as the square of the mode index. For low lying excitations, only a few Fourier modes can already produce accurate results. The PEEC method [21] expands the circuit parameters in a local basis set and thus do not have these simplifications.

(3) The local basis works, in principle, for any metallic wire network and the accuracy is determined by the mesh points chosen in a specific computation, but the physics behind the eigenmodes is not very clear. The nonlocal-basis theory is generally less computation intensive if better basis functions are chosen; it is also more "modular" and offers a better physical picture on the characteristics of the eigenmodes. Of course, for complicated structures, it is easier to pick a local basis. The wire structures are simple enough that a nonlocal basis can generally be used.

The expansion of the electric field in terms of "entire domain basis functions" had been discussed by the electrical engineering community in the past. But eventually it fell out of favor. As the reader will find out, for the class of problems discussed here, our method really works well. One of the problems in using nonlocal basis functions is that the electric field can be very different at a few places such as wire ends and junctions. We find that this problem can be overcome by introducing auxiliary variables corresponding to localized electric fields at these places. We summarize, next, the circuit equation that we shall study in different chapters in this book.

1.2 Circuit Equation

As shown in Fig. 1.1, we assume that the current density at position $\vec{r}$ inside a metallic wire structure and time t is denoted by $\vec{j}(\vec{r}, t)$, and the resistivity distribution of the system is described by $\rho(\vec{r})$, then the following equation holds

$$\rho(\vec{r})\vec{j}(\vec{r}, t) = \vec{E}_{\text{ext}}(\vec{r}, t) + \vec{E}_L(\vec{r}, t) + \vec{E}_C(\vec{r}, t). \tag{1.1}$$

This equation states that the total electric field is composed by three parts: the external field $\vec{E}_{\text{ext}}(\vec{r}, t)$, the inductive field $\vec{E}_L(\vec{r},t) = -\partial \vec{A}(\vec{r},t)/c\partial t$, and the capacitive field $\vec{E}_C(\vec{r}, t) = -\nabla\varphi(\vec{r}, t)$ where $\vec{A}(\vec{r}, t), \varphi(\vec{r}, t)$ are the vector and scalar potentials. We represent a physical quantity X as a Fourier series in time and focus on a single Fourier component: $\vec{X}(\vec{r}, t) = \vec{X}(\vec{r})\exp(i\omega t)$, with ω denoting the frequency. The inductive field can be expressed in terms of the vector potential

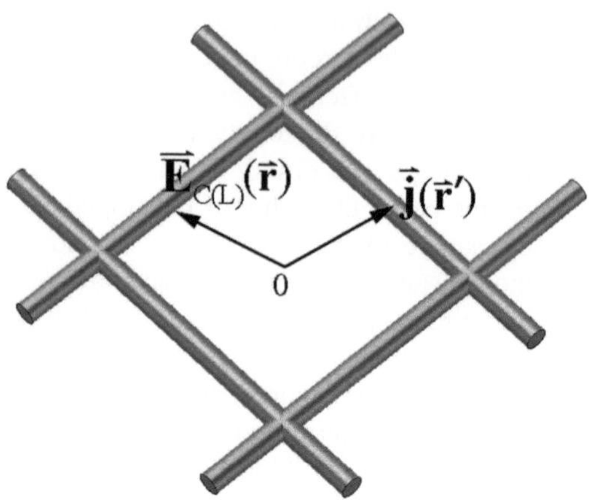

Fig. 1.1 Schematic picture of a general metallic-wire structure

$\vec{A}(\vec{r}, \omega)$ in frequency domain as: $\vec{E}_L(\vec{r}) = -\frac{1}{c}i\omega\vec{A}(\vec{r})$, where $\vec{A}(\vec{r})$ can further be expressed by the current density as

$$\vec{A}(\vec{r}) = \frac{1}{c}\int \vec{j}(\vec{r}')G(\vec{r} - \vec{r}', \omega)d^3\vec{r}' \tag{1.2}$$

in which

$$G(\vec{r} - \vec{r}', \omega) = \exp(-i\omega|\vec{r} - \vec{r}'|/c)/|\vec{r} - \vec{r}'| \tag{1.3}$$

is the retarded Green's function with c the speed of light in vacuum. The inductive field thus takes the final form as

$$\vec{E}_L(\vec{r}) = -\frac{i\omega}{c^2}\int \vec{j}(\vec{r}')G(\vec{r} - \vec{r}', \omega)d^3\vec{r}'. \tag{1.4}$$

The capacitive field can also be expressed in terms of current density as

$$\vec{E}_C(\vec{r}) = +\frac{1}{i\omega}\vec{\nabla}\int \vec{\nabla}' \cdot \vec{j}(\vec{r}')G(\vec{r} - \vec{r}', \omega)d^3\vec{r}' \tag{1.5}$$

where the charge reservation law $\nabla \cdot \vec{j}(\vec{r}, t) + \partial\rho(\vec{r}, t) = 0$ has been applied to express the charge density $\rho(\vec{r}, t)$ in terms of current density $\vec{j}(\vec{r}, t)$ in Eq. (1.5).

From Eq. (1.1), we understand that the electric fields that we are interested in are only those on the wire surface where the electric currents flow. For such fields, the dominant contributions to the integrations in Eqs. (1.4–2.5) come from those points $\vec{r}'$ that are very close to $\vec{r}$. For those points, we have $\omega|\vec{r} - \vec{r}'|c \ll 1$ and their radiation corrections are very small. Therefore, we frequently adopt the quasi-static approximation (QSA) to rewrite the Green's function as

$$G_{QS}(\vec{r} - \vec{r}', \omega) = 1/|\vec{r} - \vec{r}'| \tag{1.6}$$

and do not consider the retardation effect in the resistivity function $\rho(\vec{r})$. The QSA becomes exact in the thin-wire limit since the near-field contribution (quasi-static contribution) becomes log divergent as the wire radius turns to 0. The QSA has included correctly the most dominant contribution, the radiation corrections which are less important can be incorporated within the formulation in this book. We will come back to this point in the next chapter.

The above equations are actually the starting point of many schemes in computational EMs. For example, in the moments method Eq. (1.1) is essentially the electric field integral equation that is commonly studied [26]. A similar equation is also used in the PEEC approach. This equation will also be the central equation from which we try to understand the physics of wire structures.

Recent applications quite often focus on the response to external magnetic fields B_{ext}. Now $B = \nabla \times A$. The current can be computed from condition that the sum of the external magnetic field perpendicular to the wire and that generated by the current is zero. We obtain the relationship (the magnetic field integral equation):

$$\vec{B}_{\text{ext}}(\vec{r}) = -\int \vec{\nabla} \times \vec{j}(\vec{r}')G(\vec{r} - \vec{r}', \omega)d^3\vec{r}'. \tag{1.7}$$

For general wire structures that do not form close loops, this equation becomes essential.

This book is organized as follows: In Chaps. 2 and 3 we illustrate our approach by applying it to SRRs. The application to less symmetrical structures such as a metallic helix is described in Chap. 4. The consideration to topologically more complicated structures where three wires can meet together is described in Chap. 5. As examples the T and the H structure are described. As another more complicated structure the Jerusalem Cross is considered in Chap. 6. In the wire structures, the electric field at the end is often larger than the external field by several orders of magnitude. The behavior of wire structures under a moderate electric field is described in Chap. 7. The propagation of EM waves in a medium consisting of arrays of wire structures is described in Chaps. 8 and 9. In Chap. 8, we focus on plasmonics, whereas in Chap. 9 we focus on the magnetoelectric properties of the system.

References

1. E. Yablonovitch, Phys. Rev. Lett. **58**, 2059 (1987)
2. S. John, Phys. Rev. Lett. **58**, 2486 (1987)
3. M.S. Kushwaha, P. Halevi, L. Dobrzynski, B. Djafari-Rouhani, Phys. Rev. Lett. **71**, 2022 (1993)
4. Z. Liu, X. Zhang, Y. Mao, Y.Y. Zhu, Z. Yang, C.T. Chan, P. Sheng, Science **289**, 1734 (2000)
5. W. Wen, L. Zhou, J. Li, W. Ge, C.T. Chan, P. Sheng, Phys. Rev. Lett. **89**, 223901 (2002)
6. L. Zhou, W. Wen, C.T. Chan, P. Sheng, Appl. Phys. Lett. **82**, 1012 (2003)
7. B. Hou, H. Xie, W. Wen, P. Sheng, Phys. Rev. B **77**, 125113 (2008)

8. J.B. Pendry, A.J. Holden, D.J. Robbins, W.J. Stewart, IEEE Trans. Microw. Theory Tech. **47**, 2075 (1999)
9. T.J. Yen, W.J. Padilla, N. Fang, D.C. Vier, D.R. Smith, J.B. Pendry, D.N. Basov, X. Zhang, Science **303**, P1494 (2004)
10. D. R. Smith, W.J. Padilla, D.C. Vier, S.C. Nemat-Nasser, S. Schultz, Phys. Rev. Lett. **84**, 4184 (2000)
11. R.A. Shelby, D.R. Smith, S. Schultz, Science **292**, 77 (2001)
12. C.R. Simovski, B. Sauviac, Radio Sci. **39**, RS2014 (2004)
13. D.R. Smith, J.B. Pendry, M.C.K. Wiltshire, Science **305**, 788 (2004)
14. N. Fang, H. Lee, C. Sun, X. Zhang, Science **308**, 534 (2005)
15. T. Taubner, D. Korobkin, Y. Urzhumov, G. Shvets, R. Hillenbrand, Science **313**, 1959 (2006)
16. V.G. Veselago, Sov. Phys. Usp. **10**, 509 (1968)
17. J.B. Pendry, Phys. Rev. Lett. **85**, 3966 (2000)
18. S.T. Chui, J. Appl. Phys. **104**, 013904 (2008)
19. J.K. Gansel et al., Science **325**, 1513 (2010)
20. S.T. Chui, W.H. Wang, L. Zhou, Z.F. Lin, J. Phys. Condens. Matter **21**, 292202 (2009)
21. A.E. Ruehli, IEEE Trans. Microw. Theory Tech. **MTT-22**, 216 (1974)
22. L. Zhou, S.T. Chui, Phys. Rev. B **74**, 035419 (2006)
23. S.T. Chui, Y. Zhang, L. Zhou, J. Appl. Phys. **104**, 034305 (2008)
24. L. Zhou, S.T. Chui, Appl. Phys. Lett. **94**, 041903 (2007)
25. X. Huang, Y. Zhang, S.T. Chui, L. Zhou, Phys. Rev. B. **77**, 235105 (2008)
26. R.F. Harrington, *Field Computation by Moment Methods* (MacMillan, New York, 1968)

Chapter 2
Resonance Properties of Metallic Ring Systems: A Single Ring

2.1 Introduction

In 1968, Veselago proposed that a medium with simultaneously negative permittivity and permeability possesses a negative refractive index, and exhibits many unusual EM properties [1]. This proposal did not attract immediate attention, since it is well accepted that a natural material shows no magnetism at high frequencies [2]. A breakthrough appeared in 1999, when Pendry showed that a split-ring resonator (SRR) could provide magnetic responses at any desired frequency [3]. Metamaterials with negative refractive index were then successfully fabricated by combining SRRs and electric wires [4], and later the concept of metamaterial was greatly expanded to beyond negative-index materials. Many unusual EM phenomena were subsequently demonstrated based on metamaterials, such as negative refraction [4–10], super focusing [11–14], and subwavelength resonant cavities [15–17].

As the first realization of artificial magnetism, the SRR structures naturally attracted the most extensive attention. Circular SRRs [3, 4, 6, 18–36], rectangular SRRs [5, 37–51], SRRs with different cross-sections, metal line widths, metal thicknesses, and substrates [4–6, 19, 38, 43, 44], SRRs with different numbers of splits [19, 20, 39], and different numbers of rings [31, 47] have been studied. Applications were also proposed for the SRR, for instance, as antennas [52], as couplers for channel dropping [53]. SRRs were inserted inside a cut-off waveguide to enhance the transmission [54, 55], and as a lens for imaging [56, 57]. Recently, the designs and fabrications of isotropic metamaterials began to draw intensive attentions [58–64].

Many theoretical efforts were devoted to understand the exotic EM wave properties of the SRR structures. Pendry et al. [3] first analyzed the resonance properties of a SRR by assuming a metallic ring as a single lumped element with empirical circuit characteristics. Later, Shamonin et al. considered more inductive/capacitive effects by assuming the SRR to consist of an infinite number of lumped circuit elements [25]. Many other analytical methods were developed to study the

S. T. Chui and L. Zhou, *Electromagnetic Behaviour of Metallic Wire Structures*,
DOI: 10.1007/978-1-4471-4159-4_2, © Springer-Verlag London 2013

properties of SRRs [24, 26–29, 31, 32, 47, 48] from various aspects. However, in all these approaches, the inductive/capacitive effects in the rings were not *completely* considered, since self (mutual)-inductive/capacitive effects exist everywhere inside the system. In addition, all these approaches need a set of empirical circuit parameters that cannot be calculated rigorously. Those empirical parameters were generally determined under some approximations [3, 25]. The SRR systems have also been studied by numerical calculations [37, 38, 44, 45, 49, 50]. Although such full-wave studies contain all relevant field information, it is sometimes difficult to extract useful properties of an SRR, such as the bianisotropy polarizabilities, from the obtained information.

We recently developed an analytical approach on more rigorous ground for metallic wire systems [33–36]. In this chapter, we illustrate our approach to wires in ring geometries, and apply it to quantitatively study the EM resonance properties of a single ring SRR [33]. In our theory, the inductive/capacitive effects are *fully* included and all circuit parameters are calculated *rigorously*. One of the key ideas of our approach is a simple way to implement the boundary condition that the current is zero at the free ends of the wire. This idea consists of the introduction of localized electric fields at ends and junctions. We show that the circuit equations can be analytically solved in the thin-wire limit, leading to useful analytical formulas. Our theoretical results were all successfully verified by FDTD simulations on realistic systems and/or available experiments. This chapter is organized as follows. We describe our simple picture of split rings in the next section, briefly review the theoretical developments in Sects. 2.3 and 2.4, and then apply our theory to study the EM resonance properties of a single ring SRR analytically (Sect. 2.5) and numerically (Sect. 2.6). For those readers who prefer to go directly to the heart of the matter, they can start with Eq. (2.18) where the circuit equation in the angular momentum basis including the localized field is introduced. We summarize our results in the last section.

2.2 Simple Physical Picture

Before we start going into the details of dealing with the problem quantitatively, first we would like to explain the simple physics for this class of structures. Central to the understanding of the EM response is the resonance mode of a wire structure. These resonance modes are characterized by the spatial dependence of the tangential current along the wire. To a good approximation the current depends on the *arc length* of the wire sinusoidally with a period that is twice the length of the wire divided by an integer. This is independent of the shape of the wire. The more spatial oscillation the current exhibits, the higher the resonance frequency of the normal mode. We shall see why this is true and how this will be modified quantitatively when the wire becomes thicker.

From this spatial dependence of the current one can calculate the response of the wire to external EM fields. The response is characterized by the multipole

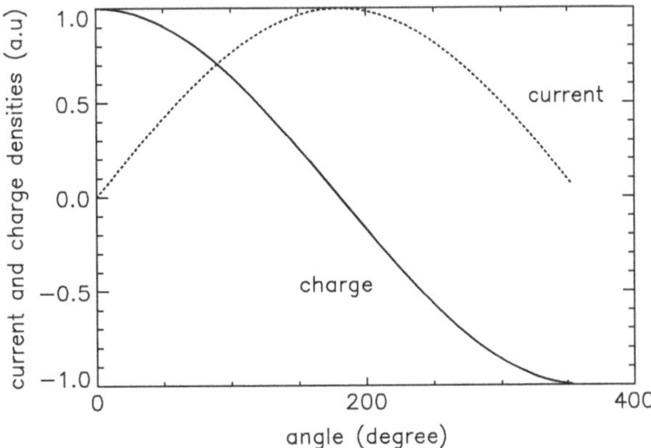

Fig. 2.1 The current and charge densities as a function of angle for the lowest resonance of a single ring SRR

moments induced. Because different wire structures can have different shapes, the multipole moments will depend on the shape of the wire structure. The magnetic moments depend on the current and can be easily calculated. For example, for the split ring we get a finite magnetic moment, whereas for the dipole antenna we do not, even though the normal modes, when expressed in arc length, are very similar. The electric multipole moments can be derived from the charge density ρ, which can be determined from the charge-current conservation law as $\rho = i\nabla \cdot \vec{j}/\omega$. To illustrate, we show in Fig. 2.1 the dependence of the current on the azimuthal angle for the lowest mode of a split ring. Note that the current is zero at the free ends. The corresponding charge density is also shown in Fig. 2.1. For this mode, the electric dipole moment is in the plane of the ring, whereas the magnetic dipole moment is perpendicular to the ring. Either an external electric or magnetic field can excite this normal mode, generating both an electric, and a magnetic moment. This picture remains valid as the number of wires is increased or the topology is changed.

2.3 Circuit Parameters for a Single Ring

We consider a single ring of radius R with a small gap at $\phi = 0°$ lying on the xy-plane, as shown in Fig. 2.2 [33]. In what follows, a common time-varying factor $e^{i\omega t}$ is omitted for every quantity. We assume that $a \ll R$, where a is the radius of the metal wire forming the ring. For a good metal, the skin effect dictates that the current should mainly distribute on the metal surface, within a thin layer of a thickness equal to the metal skin depth $\delta = (\mu\omega\sigma/2)^{-1/2}$. For typical material like Cu with a resistivity $1/\sigma = 1.68 \times 10^{-8}\,\Omega \cdot m$, we get

$$\delta = 3.76\lambda^{1/2}\,\mu m \qquad (2.0)$$

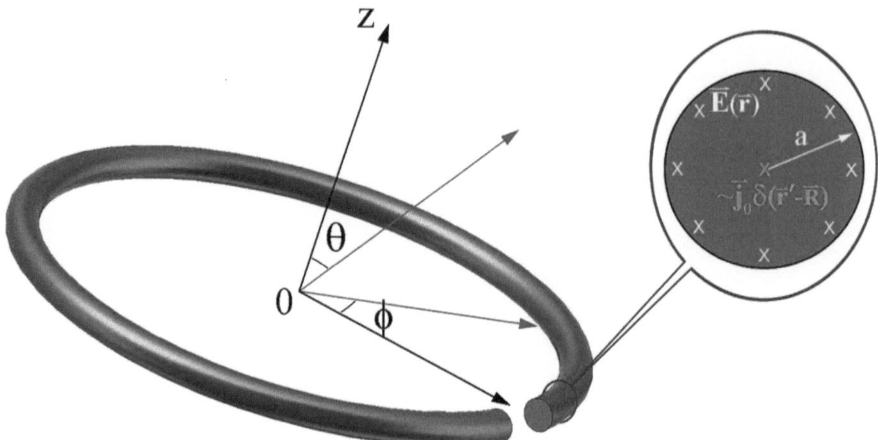

Fig. 2.2 Geometries of a single ring SRR

at a wavelength of λ measured in m. Typically, the largest wavelength of interest is of the order of twice the circumference: $\lambda \approx 4\pi R$ (see below). We thus get $\delta/R \approx 0.47 \times 10^{-4}/\lambda^{1/2}$. From microwave to infrared frequencies with the wavelengths of the order of a centimeter to a micron, the current flows on the conductor's outer surface within a layer of thickness of the order of the skin depth, so that $\vec{j}(\vec{r}')$ is basically a very complicated function of $\vec{r}'$. However, when the wire is very thin, we can simplify the realistic current distribution as a delta-function localized in the middle of the wire,

$$\vec{j}(\vec{r}') = \vec{e}_{\phi'} I(\phi') \sin\theta' \delta(\cos\theta') \delta(r' - R)/R \tag{2.1}$$

With $\vec{e}_{\phi'} = -\sin\phi'\vec{e}_x + \cos\phi'\vec{e}_y$. This simplification will *not* generate any significant errors for calculating the fields outside the metal wire in the thin-wire limit. On the other hand, both the inductive $\vec{E}_L(\vec{r})$ [Eq. (1.4)] and the capacitive electric fields $\vec{E}_C(\vec{r})$ [Eq. (1.5)] should still be calculated on the outer surface of the metal wire where the current *physically* flows. Again considering the fact $a \ll R$, we understand that $\vec{E}_L(\vec{r})$ and $\vec{E}_C(\vec{r})$ would not vary dramatically around the wire (as long as on the same position of the ring with a fixed ϕ), so that we can pick up a particular (convenient) point on the wire surface to calculate these fields.

The particular symmetry of the ring geometry indicates that all physical quantity (i.e., $I(\phi), \vec{E}_L(\vec{r})$, and $\vec{E}_C(\vec{r})$, etc.) are periodic functions of ϕ with period of 2π. Therefore, we can expand those quantifiers as a Fourier series of the azimuthal angle ϕ. Such a choice of basis will simplify the calculations dramatically, as will be shown soon. Substituting Eq. (2.1) into Eq. (1.4), we can choose a particular observation point on the wire surface to calculate the inductive field *projected along the wire direction,*

$$\vec{E}_L(\phi) \cdot \hat{e}_\phi = -i\omega \int (\hat{e}_\phi \cdot \hat{e}_{\phi'})I(\phi')g(\phi,\phi')Rd\phi'/c^2, \tag{2.2}$$

where the reduced Green's function is defined by

$$g(\phi,\phi') = G(\vec{r},\vec{r}')|_{r'=R,\theta'=\pi/2;r=R-a,\theta=\pi/2} \tag{2.3}$$

so that r is at the surface and r' is at the center of the wire. Here the observation point is selected at $(r = R - a \quad \theta = \pi/2, \quad \phi = \phi)$ for the convenience of calculations and $r' = R, \theta' = \pi/2$ come from the fact that current is localized at the wire center. Other choice of observation point will not affect the final results under the thin-wire limit, but the calculations may not be as easy as the present choice.

Putting the Fourier expansion $I(\phi') = \sum_{m=-\infty}^{+\infty} I_m e^{im\phi'}$ into Eq. (2.2), we obtain that

$$E_L^m = -i\omega \sum_{m'} L_{mm'} I_{m'} \tag{2.4}$$

where

$$E_L^m = \frac{1}{2\pi} \int_0^{2\pi} (\vec{E}_L \cdot \vec{e}_\phi)e^{-im\phi}d\phi \tag{2.5}$$

and

$$L_{mm'} = \frac{1}{2\pi c^2} \iint (\hat{e}_\phi \cdot \hat{e}_{\phi'})e^{i(m'\phi'-m\phi)}g(\phi,\phi')Rd\phi'd\phi \tag{2.6}$$

are the elements of the inductance matrix. Similarly, we obtain the result

$$E_C^m = -\frac{1}{i\omega} \sum_{m'} (C^{-1})_{mm'} I_{m'}, \tag{2.7}$$

where

$$E_C^m = \frac{1}{2\pi} \int_0^{2\pi} (\vec{E}_C \cdot \vec{e}_\phi)e^{-im\phi}d\phi \tag{2.8}$$

is the Fourier component of the capacitive field projected along the wire direction, and

$$(C^{-1})_{mm'} = -\frac{1}{2\pi} \iint (im')e^{i(m'\phi'-m\phi)} \frac{1}{(R-a)} \left(\frac{\partial}{\partial\phi}g(\phi,\phi') \right) d\phi'd\phi \tag{2.9}$$

are the capacitance matrix elements. Both the inductance and capacitance matrixes can be calculated numerically by performing the integrations in Eqs. (2.6) and (2.9). For the present symmetric ring geometry, analytical formulas can be derived, which offer us physical insights. The details of this are described in Appendix in this chapter. The result under the quasi-static approximation (QSA) contains the leading log contribution and will be discussed here. Corrections beyond the QSA are discussed in the Appendix. Putting the formula ($r_>$, $r_<$ are the larger and the smaller of r, r')

$$\frac{1}{|\vec{r} - \vec{r}'|} = \sum_{l=0}^{\infty} \sum_{m=-l}^{l} \frac{4\pi}{2l+1} Y_{lm}(\theta, \phi) Y_{lm}^*(\theta', \phi') \frac{r_<^l}{r_>^{l+1}} \tag{2.10}$$

into Eqs. (2.6) and (2.3), we find for the present ring geometry that $L_{mm'}$ and $(C^{-1})_{mm'}$ are diagonal with matrix elements given by

$$L_{mm'} = \delta_{mm'} L_m = \delta_{mm'}(A_{m-1} + A_{m+1})\pi/c^2$$
$$(C^{-1})_{mm'} = \delta_{mm'}(C_m)^{-1} = \delta_{mm'}m^2 A_m 2\pi/[R(R-a)] \tag{2.11}$$

The function A_m is defined as

$$A_m = \sum_{l=|m|}^{\infty} \frac{(l-m)!}{(l+m)!} \alpha^l \left[P_l^m(0)\right]^2 \tag{2.11a}$$

where P_l^m is the associated Legendre function, and $\alpha = (R-a)/R < 1$.

There are several important points about the circuit parameters:

(1) In the thin-wire limit ($a \ll R$), the sum in Eq. (2.6) can be carried out analytically. We found (see Appendix in this chapter) an asymptotic form

$$A_m \approx D_m - \ln(2a/R)/(\alpha^m \pi). \tag{2.12}$$

Thus in this limit all the A_m are the same. The inductances are independent of m. We obtain

$$L_m = -2\ln(2a/R)/c^2. \tag{2.12a}$$

Similarly we get

$$1/C_m = -2m^2 \ln(2a/R)/R^2. \tag{2.12b}$$

This logarithmic divergence is a typical characteristic of a thin-wire system [65]. It comes out because of the factor of $1/|\vec{r} - \vec{r}'|$ in the Green's function inside the integrations for the inductive/capacitive fields. Such logarithmic dependence of circuit parameters on wire radius is the basis to justify for the QSA, as we have discussed in Chap. 1. In wire structures of lower symmetry, the circuit parameters are no longer "diagonal" and we have to consider the "off-diagonal" elements of matrixes L and C^{-1}. As we shall show in the next chapter this off-diagonal element with m not equal to m' is not log divergent and much smaller. Much of computational electromagnetics involves calculating the inverse of the matrix $L_{mm'}$ and/or $(C^{-1})_{mm'}$ in a local basis. In our basis, it is nearly diagonal; the inversion of the matrixes $L_{mm'}$ and $(C^{-1})_{mm'}$ becomes simpler, the physical implication of the results is much easier to understand.

(2) In our basis $(C_m)^{-1}$ is proportional to m^2. Thus I_m, which is of the order of $C_m E_m$, decreases rapidly as m is increased. This sets the limit of the number of Fourier mode that needs to be retained. In the following, to emphasize this point, we sometimes write $(C_m)^{-1}$ as $m^2/\overline{C_m}$.
(3) One can consider the $m = 0$ components to correspond to the lump circuit elements normally considered. The approach here thus provides for a simple extension of previous considerations. However, there is no $m = 0$ capacitive term. To include a capacitive effect, the $m = 1$ term need to be included.
(4) If one goes beyond the QSA, a damping term proportional to kR will be introduced in addition to the log divergent terms, as is discussed in the Appendix.

The above characteristics are true for all wire structures and will be exploited in our discussion in later chapters.

2.4 Implementation of the Boundary Conditions

We can now solve the circuit Eq. (1.1)

$$\rho(\vec{r})\vec{j}(\vec{r},\, t) = \vec{E}_{\text{ext}}(\vec{r},\, t) + \vec{E}_L(\vec{r},\, t) + \vec{E}_C(\vec{r},\, t).$$

in the basis that we have chosen. So far we have assumed that the Fourier components of the current are independent variables. There is a boundary condition that the current at the gap vanishes, $I(\phi = 0) = 0$, which can be written as $\sum_m I_m = 0$ in terms of the Fourier components. We describe, in this chapter, two ways that we have tried to implement this condition. They lead to identical results, thus providing confidence in our method. We describe these next.

2.4.1 The Gap Resistance Approach

The boundary condition of zero current can be incorporated by introducing a very large resistance in the gap region. In terms of the Fourier components and using the formulas derived in last section, Eq. (1.1) can be written as the following matrix equation,

$$\sum_{m'} H_{mm'} I_{m'} = E^m_{\text{ext}}, \tag{2.13}$$

where

$$H_{mm'} = \bar{\rho}(m - m') + i\omega L_m (1 - \Omega_m^2/\omega^2)\delta_{mm'}. \tag{2.14}$$

Here, $\bar{\rho}(m - m')$ is the Fourier component of the normalized resistivity function defined by $\bar{\rho}(\phi) = \rho(\phi)/S$ (S is the wire cross-section area and $\rho(\phi)$ is the true resistivity function) and $\Omega_m = 1/\sqrt{L_m C_m}$. In the thin-wire limit ($a/R \to 0$), $\Omega_m \to m\omega_u$ where

$$\omega_u = c/R \qquad (2.14a)$$

is the frequency unit of the present problem. Suppose $\bar{\rho}(\phi) = \bar{\rho}_0$ inside the gap of width Δ centered at $\phi = 0$ and $\bar{\rho}(\phi) = r_c$ elsewhere (in the metallic wire), we found easily that

$$\bar{\rho}(m - m') = r_c \delta_{m,m'} + \frac{\sin[(m - m')\Delta/2]}{\pi(m - m')}(\bar{\rho}_0 - r_c). \qquad (2.15)$$

Putting Eq. (2.15) into Eq. (2.14) and then diagonalizing the H matrix, we obtain the eigenvalues $\{\lambda_j\}$ and eigenvectors of all the EM modes, from which we can calculate the resonance frequencies, induced EM dipole moments, and the polarizabilities as well as the bianisotropic polarizabilities. We will present numerical solutions of Eq. (2.13) for several examples in the following sections.

2.4.2 The Local Field Approach

The matrix problem (2.13) can be analytically solved under some reasonable assumptions, which help us to reach at an alternative approach to implement the boundary condition at the wire ends. In the limit of $\Delta \to 0$, we found from Eq. (2.15) that $\bar{\rho}(m - m') \to r_c \delta_{m,m'} + r$, where $r = \Delta \bar{\rho}_0/2\pi \gg r_c$ is proportional to the total resistance across the gap. We put $\bar{\rho}(m - m')$ into Eq. (2.14) and rewrite the vector-matrix equation as

$$HI = E_{\text{ext}} \qquad (2.16)$$

with $H = H_0 + X$, where $H_0 = rM$, in which $M_{i,j} = 1$, and $X = diag[X_m]$ is a diagonal impedance vector-matrix with elements defined as

$$X_m = r_c + i[L_m \omega - m^2/(\overline{C}_m \omega)]. \qquad (2.17)$$

We find that the vector-matrix problem (2.16) can be analytically solved in the limit of $r \to \infty$, as follows. Now $H_0 I = 0$ so long as $\sum_m I_m = 0$. This current distribution is such that its magnitude is zero at the gap, which is a solution satisfying the desired boundary condition. However, not all I satisfying $\sum_m I_m = 0$ are the correct solutions of the circuit equation Eq. (2.13) because the internal emf inside the ring is not generally zero: $E_{\text{int}}(\phi) = \sum_m E_m \exp(im\phi) \neq 0$, where $E_m = X_m I_m - E_m^{\text{ext}}$. The only way the circuit equation Eq. (1.1) can be satisfied is if $E_{\text{int}} = \sigma\delta(\phi)$ for some constant σ; the internal emf is zero inside the ring except at the gap where it is counterbalanced by the infinite gap resistance r. For this to be true, it is necessary that all E_m be the same. We thus write

$$E_m = X_m I_m - E_m^{\text{ext}} = \sigma \qquad (2.18)$$

so that $I_m = (E_m^{\text{ext}} + \sigma)/X_m$. The existence of these fields σ can also be appreciated from the law of charge conservation: $\partial\rho/\partial t = -\nabla \cdot \vec{j}$. At the end, the net divergence of the current $\vec{j}$ is not zero. We thus expect localized time-varying charges and thus localized electric fields, which is represented by σ. Since $\sum_m I_m = 0$, we find the solution of σ given by:

$$\sigma = -\frac{\sum_m E_m^{\text{ext}}/X_m}{\sum_m 1/X_m} \qquad (2.19)$$

The resonance frequency ω_c is determined from the condition that a nontrivial solution still exist even when $E_{\text{ext}} = 0$. We thus arrive at a sufficient resonance condition:

$$\sum_m 1/X_m(\omega_c) = 0. \qquad (2.19a)$$

As we shall see, this applies to modes of even symmetry. There are additional trivial resonances of odd symmetry.

Substituting in the expression for X_m, we arrive at the equation that determines the entire spectrum:

$$1 + \sum_{m=1}^{\infty} 2(L_0\omega^2 - ir_c)/(L_m\omega^2 - m^2/\overline{C}_m - ir_c) = 0.$$

Because X_{m0} is proportional to m^2, only a few terms in m need to be included in the above sum. We have tested the eigenfrequencies obtained from this equation with results from numerical FDTD calculations and found good agreement. Substituting in the value of the end field σ into the circuit equation, we obtain the expression for the current response to the external field:

$$I_m = (E_{\text{ext}}^m - \sum_p E_{\text{ext}}^p/X_p/\sum_n 1/X_n)/X_m. \qquad (2.19b)$$

The scattered field can then be obtained from the sum $E_L(r) + E_C(r)$ for r outside of the wire.

The above consideration motivates the introduction of the boundary electric field σ. For more complicated circuit structures with free ends, it turns out to be much easier to directly deal with the boundary electric field, since it is difficult to introduce the infinite boundary resistance as we did for the split ring. Instead of imposing zero boundary current constraint directly, we introduce the idea that there is a new additional variable corresponding to a localized electric field σ at the free end of the split ring. The value of σ is chosen to satisfy the desired boundary condition—the current at the end should vanish. This boundary field is, in some sense, like a Lagrange multipler. The circuit equation, in terms of the Fourier component of the impedance, the external electric field E_{ext}^m and the localized electric field at the free ends, σ, is then given by

$$X_m I_m = E_{\text{ext}}^m + \sigma \tag{2.20}$$

which is consistent with Eq. (2.18). Solving Eq. (2.10) to obtain $I_m = (E_{\text{ext}}^m + \sigma)/X_m$, and then imposing the boundary condition $\sum_m I_m = 0$, we reach again at Eq. (2.19) for the solution of σ. Such consistency justifies the local field approach to implement the boundary conditions at the wire ends.

In summary, we have completed the theoretical development of the mode-expansion theory for a single-ring system, arriving at the circuit equation Eq. (2.13). This is the basic matrix equation to be solved. The circuit parameters are explicitly given in Sect. 2.3 and the boundary conditions are implemented in Sect. 2.4. In what follows, we shall apply this theory to study a single ring SRR illustrated in Fig. 2.2. For such a structure, the circuit matrix Eq. (2.13) can be analytically solved under some particular assumptions, leading to analytical results on both resonance frequencies and current distributions. Alternatively, in general situations where the analytical approach does not apply to, we can also solve Eq. (2.13) numerically to obtain all the necessary information. These two approaches are complementary to each other, and combining them significantly deepens our understandings on the inherent physics of the problem, and thus forms a comprehensive picture on the physical problem. In the following, we will separately describe the analytical and numerical approaches to study the single ring SRR.

2.5 Analytical Results for a Single Ring SRR

The circuit problem of a single ring SRR (Eq. (2.13)) can be solved analytically under the following limits: (1) $\Delta \to 0$; (2) $a/R \to 0$; (3) the metal is perfect ($r_c = 0$) and the gap is an ideal insulator ($r \to \infty$). Under these conditions, we can *rigorously* solve the vector-matrix problem which now takes the form

$$HI = 0 \tag{2.21}$$

with the Hamiltonian vector-matrix H defined in last section. The resonance frequency ω_c is determined from the condition that a nontrivial solution still exist even when $E_{\text{ext}} = 0$. We have thus set $E_{\text{ext}} = 0$ here since we only need the information of the resonance eigenmodes.

Because of the reflection (m to $-m$) symmetry of X, $X_m = X_{-m}$ there are two classes of solutions for Eq. (2.21), corresponding to even and odd symmetries under the transformation from m to $-m$. Those with odd symmetries (i.e., $I_{-m} = -I_m$) are purely geometrical resonance modes with resonance frequencies determined by $X_m = 0$. Explicitly, we found from Eq. (2.17) that the resonance frequencies are given by

$$\omega_{2m} = m\omega_u, \quad m = 1, 2, \ldots \tag{2.22}$$

We note that these modes do not depend on the gap resistance r. The odd symmetry modes are simple in the limit of narrow gap $\Delta \to 0$ because the condition that $I(\phi = \Delta/2) = -I(\phi = -\Delta/2)$ implies that the current is automatically zero at the gap and there is no additional need to deal with this constraint. As the gap widens, this is no longer true. This limit can be studied with extension of the ideas described below as is described in later chapters.

To study another set of resonance modes with even symmetries (i.e., $I_{-m} = I_m$), we describe a formal approach to solve the circuit Eq. (2.21), which can be easily extended to more complicated situations. This approach explicitly displays the eigenvectors, which can be used to calculate the responses of the structures to external fields. Because r is very big, we solve the matrix problem (2.21) by a standard perturbation method. For the unperturbed matrix problem, $\mathbf{H_0}\mathbf{I}^{(0)} = \mathbf{0}$, any vector $\mathbf{I}^{(0)}$ satisfying

$$\sum_m I_m^{(0)} = 0 \tag{2.23}$$

is a solution. Now consider the full matrix problem Eq. (2.21). Assume the solution to Eq. (2.21) can be written as $\mathbf{I} = \mathbf{I}^{(0)} + \mathbf{I}^{(1)}$, where $\mathbf{I}^{(1)}$ is of the order of $r^{-1}\mathbf{I}^{(0)}$. Put $\mathbf{I}$ into Eq. (2.23), since $\mathbf{H_0}\mathbf{I}^{(0)} = \mathbf{0}$, we get $\mathbf{HI} = (r\mathbf{M} + \mathbf{X})(\mathbf{I}^{(0)} + \mathbf{I}^{(1)}) = r\mathbf{MI}^{(1)} + \mathbf{XI}^{(0)} + o(r^{-1}) = 0$. Substituting in the elements of $\mathbf{X}$ and $\mathbf{M}$, we get $I_m^{(0)} = E'/X_m$ where $E' = -r\sum_{m'} I_{m'}^{(1)}$ is a constant independent of m. Employing the constrain (2.23), we arrive at the Eq. (2.19a) : $\sum_m 1/X_m = 0$, which leads to the following polynomial equation

$$1 + \sum_{m=1}^{\infty} \frac{2L_0\omega^2}{L_m\omega^2 - 1/C_m} = 0 \tag{2.24}$$

to determine the resonance frequencies. Choosing an appropriate normalization constant K, in the limit $1/r \to 0$, we find the eigenvector at resonance to be

$$\mathbf{I} = \mathbf{I}^{(0)} = K[\ldots, 1/X_2, 1/X_1, 1/X_0, 1/X_1, 1/X_2, \ldots]^T. \tag{2.25}$$

We emphasize that Eqs. (2.24) and (2.25) are the ***exact*** solutions of the matrix problem (2.21), since the perturbation theory becomes ***exact*** in the limit of $1/r \to 0$.

We now solve Eq. (2.24) analytically. Since as $a/R \to 0$, we have Eq. (2.12): $L_m \to \ln(R/a)$ and $1/C_m \to m^2 \ln(R/a)$, using the identity $k\pi \cot(k\pi) = 1 + \sum_{m=1}^{\infty} 2k^2/(k^2 - m^2)$ [67], we found that Eq. (2.24) can be written as $(\omega/\omega_u) \cot(\omega\pi/\omega_u) = 0$, leading to the following solutions

$$\omega_{2m+1} = (m + 1/2)\omega_u, m = 0, 1, 2, \ldots \tag{2.26}$$

In terms of the wavelength, this condition is equivalent to $\lambda_{2m+1} = 4\pi R/(2m+1)$.

We next study the current distribution for the resonance modes. In the thin-wire limit, we get $X_m = iL_m[\omega^2 - m^2/(L_m \overline{C}_m)]/\omega = (iL\omega_u^2/\omega)[(\omega/\omega_u)^2 - m^2]$. Therefore, according to Eq. (2.25), the current in real space is given by $I(\phi) = \sum_m e^{im\phi} I_m = K \sum_m e^{im\phi}/X_m = -iK\omega/(L\omega_u^2) \sum_m e^{im\phi}/[(\omega/\omega_u)^2 - m^2]$. Using the identity $\cos k(\phi + \pi) = k \sin k\pi [\sum_m e^{im\phi}/(k^2 - m^2)]/\pi$, we obtain, for the nth even resonance mode,

$$I(\phi) = -iK/(L\omega_u^2)\pi \sin(n + 1/2)\phi. \qquad (2.26a)$$

In terms of the arc length $s = R\phi$ and the circumference $l = 2\pi R$. This eigenfunction can also be written as $I(s) \propto \sin \pi s (2n + 1)/l$, the same as what one expects for a straight wire. Thus, in terms of the arc length, the nature of the eigenstate does not depend much on the shape of the wire, as we discussed in Sect. 2.2.

We next discuss the response of the wire structure to external fields. The discussion in this section is general and not restricted to split rings. We shall use the notation of linear algebra so that the external electric field is written as $|E>$. In Eq. (2.16) $\mathbf{H}$ is a sum of a big ($\mathbf{H_0}$) and a small term ($\mathbf{X}$). We try to calculate the eigenstate $|f>$ by perturbation theory as $|f>=|w_f>+|e_f>$ where $H_0|w_f> = 0$ and $|e_f>$ is the perturbative correction. From the condition that $\mathbf{H}|f>=0$ at the resonance frequency, we obtain

$$H_0|f> \ = H_0|e_f> \ = -X(\omega_f)|f>$$

In the language of the discussion above, $H_0|e_f> \ = -\sigma_f$, the localized electric field at the ends for the fth eigenstate. Under an external field E and at an arbitrary frequency, we expand the current in terms of the eigenfunctions as $I = \sum_j I_f|f>$ and get

$$HI \approx \sum_f I_f[H_0 + X(\omega)]|f> \ = \sum_f I_f[X(\omega) - X(\omega_f)]|f> \ = |E>$$

We thus get

$$\sum_f I_f <g|[X(\omega) - X(\omega_f)]|f> \ = \ <g|E> \qquad (2.27)$$

where the notation $<f|g> \ = \int fg$ is used. Solving this matrix equation, we obtain the coefficients I_f.

For frequencies close to the lowest mode $|0>$, we include only the term with $f = 0$ in the summation on the right hand side of the above equation. We get approximately

$$I_0 = \ <0|E> \ / <0|[X(\omega) - X(\omega_f)]|0> \qquad (2.27a)$$

From these and from Eq. (2.26) the induced electric and magnetic dipole moments **P** and **M** induced by external electric or magnetic fields can be computed.

The information of the eigenvectors enables us to evaluate the EM responses of the system. At the nth resonance mode, from Eqs. (2.1) and (2.27), we get the current density given by $\vec{j}(r') = \vec{e}_\phi K \sin[(n + 1/2)\phi] \sin \theta \delta(\cos \theta)\delta(r' - R)/R$. The corresponding charge density is given by $\rho(r') = -i\nabla \cdot \vec{j}/\omega = -iK(n + 1/2)$ $\cos[(n + 1/2)\phi] \sin \theta \delta(\cos \theta)\delta(r' - R)/\omega R^2$. For example, for n $= 0, j \propto \sin \phi/2$, $\rho \propto \cos \phi/2$, as is illustrated in Fig. 2.1

2.6 Numerical Results for a Single Ring SRR

Away from the thin-wire limit when the three conditions mentioned in Sect. 2.5 are not satisfied, the circuit equation for such systems can be solved numerically. We discuss this in this section. One main lesson we learn is that not that many Fourier modes need to be included. The convergence is very fast. This comes about because the inverse capacitance and hence the impedance X_m increases rapidly as m^2 for increasing m.

In the microwave frequency regime, we can safely set the metal's resistivity to zero (i.e., $r_c = 0$), and assume a very large but finite r for the gap resistance. We found from numerical calculations that the final results do not depend on r when $r \rightarrow \infty$. For a particular single ring SRR characterized by the wire radius a/R and the air gap size Δ, we can unambiguously compute all the circuit parameters (i.e.., L_m and C_m) and thus the matrix elements $H_{m,m'}$, based on the formulas developed in Sect. 2.3. We then diagonalize the H matrix through

$$\tilde{H} = P^{-1}HP, \tilde{I} = P^{-1}I, \tilde{E}_{ext} = P^{-1}E_{\text{ext}}, \tag{2.28}$$

where P is the transformation matrix containing the eigenvectors of the **H** matrix, and get from Eq. (2.13) that,

$$\tilde{I}_m = \frac{\tilde{E}^m_{ext}}{\lambda_m(\omega)} \tag{2.29}$$

where λ_m is the mth eigenvalue of the **H** matrix. Therefore, the resonance frequencies of the system are determined by the condition $\lambda_m(\omega) \rightarrow 0$. Numerically, we determine the resonance frequencies by the condition that the magnitude of the lowest eigenvalue of the matrix H exhibits a minimum.

The magnitude of the lowest eigenvalue of matrix **H** is shown as a function of frequency in Fig. 2.3 for a typical single ring SRR with $\Delta = \pi/40$ and $\alpha = 0.99$. The general agreements among different sets of calculations show that the adopted approximations, namely taking finite values of angular momentum cut off M_{max} and gap resistivity parameter r, do not introduce any significant errors. The series

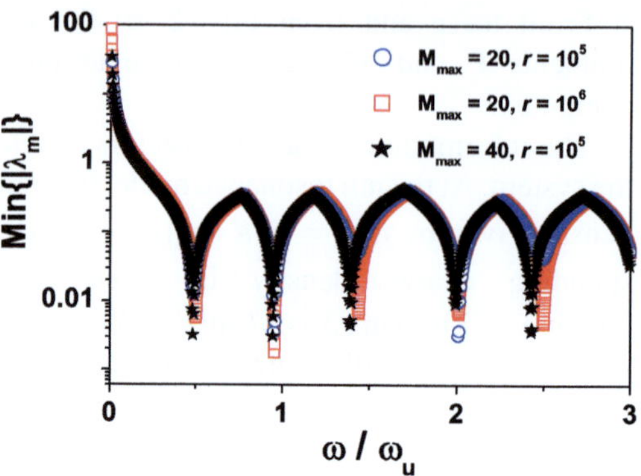

Fig. 2.3 $\min[|\lambda_m|]$ as functions of ω/ω_u for a single-ring SRR with $\Delta = \pi/40$ and $\alpha = 0.99$, calculated with different values of $M_{\max}$ (the cutoff value of m) and r (in units of $\mu_0\omega_u$). (From Ref. [33])

of resonances in Fig. 2.3 can be categorized into two classes. The even-numbered resonances ω_{2m} coincide well with the intrinsic resonances Ω_m, which, in consistency with the analytical results Eq. (2.22), become $\Omega_m \to m\omega_u$ in the thin-wire limit $(a/R \to 0)$. Since $\lambda_{2m} = 2\pi c/\omega_{2m} = m \cdot 2\pi R$, these resonances are solely determined by the ring geometry, with currents forming standard standing waves in the ring. On the other hand, the resonance frequencies of the odd-numbered eigenmodes also match with the analytical results Eq. (2.26) very well.

The eigenvectors (contained in the matrix P defined in Eq. (2.28)) are shown in Fig. 2.4 for the lowest four resonances. The odd-numbered (even-numbered) resonance modes possess symmetrical (anti-symmetrical) eigenvectors with respect to index m, which is consistent with the analytical results. In addition, for the even-numbered modes, we note that only the $\pm m$ Fourier components are excited for the $2m$th resonance eigenmode, verifying the analytical predictions. In contrast, for the odd-numbered modes, all Fourier components are excited but the contribution a Fourier component decrease significantly as m increases, again in agreement with the analytical results.

With the current distribution ($\{I_m\}$ and in turn $\vec{j}(\vec{r})$) explicitly known by diagonalizing the **H** matrix, we can compute the electric and magnetic dipole moments **P** and **M** induced by external electric or magnetic fields by

$$M = \frac{1}{2c}\int \left(\vec{r}\times\vec{j}\right)d\vec{r}, \quad P = \frac{i}{\omega}\int d\vec{r}(\nabla \cdot \vec{j})\vec{r} \tag{2.30}$$

Explicitly, we find the SRR to possess the following nonzero components of dipole moments:

$$\begin{cases} p_x = \sum_j \frac{1}{\lambda_j(\omega)}\frac{\pi R}{\omega}\left[P_{-1j} - P_{1j}\right]\cdot\tilde{E}_{ext}^j \\[2mm] p_y = \sum_j \frac{1}{\lambda_j(\omega)}\frac{\pi R}{i\omega}\left[P_{-1j} + P_{1j}\right]\cdot\tilde{E}_{ext}^j \\[2mm] m_z = \sum_j \frac{\pi R^2}{\lambda_j(\omega)}P_{0j}\cdot\tilde{E}_{ext}^j \end{cases} \tag{2.31}$$

Fig. 2.4 The eigenvector distributions for the lowest eigenmode at the lowest four resonance frequencies. (From Ref. [33])

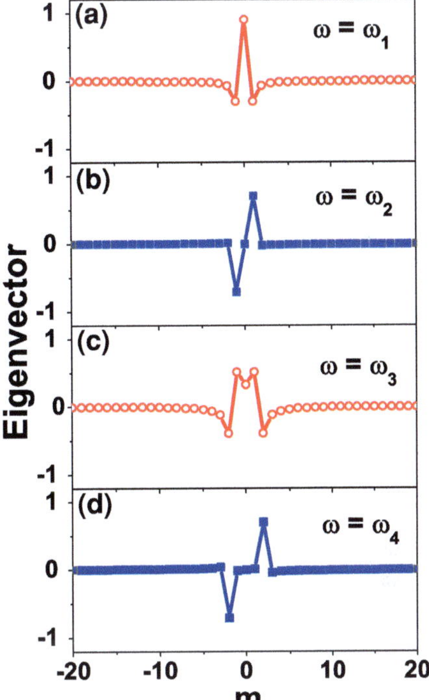

where

$$\tilde{E}^{j}_{ext} = \sum_{l} (P^{-1})_{jl} E^{l}_{ext} = \frac{1}{2\pi} \int \sum_{l} (P^{-1})_{jl} \vec{E}_{ext}(\vec{r}) \cdot \vec{e}_{\phi} \exp(-il\phi) d\phi \qquad (2.32)$$

represents the external field component projected on the jth eigenmode. A common feature of metallic wire structures is that quite often either an external magnetic or electric field alone can induce both electric and magnetic dipole moments. This property is called magnetoelectric or bianisotropic, which we will discuss explicitly in the following.

Consider different probing plane waves with (a)$\vec{E}\|\hat{y}, \vec{k}\|\hat{z}$, (b)$\vec{E}\|\hat{y}, \vec{k}\|\hat{x}$, (c)$\vec{E}\|\hat{x}, \vec{k}\|\hat{z}$, (d)$\vec{E}\|\hat{x}, \vec{k}\|\hat{y}$, we show in Fig. 2.5 the induced dipole moments as functions of the frequency. The odd-numbered resonance modes possess both magnetic (m_z) and electric (p_y) responses, while the even-numbered ones exhibit only electric (p_x) responses. The appearance of p_y is always accompanied by the appearance of m_z, which manifests the bianisotropy property of the SRR [26, 30, 32]. Symmetry restricts a probing field to excite only a particular set of resonance modes of the SRR, and therefore, not all the resonance peaks appear simultaneously in each spectrum. We have performed FDTD simulations [66] to verify these theoretical results. To model a single ring SRR with $R = 4$mm, $a = 0.1$mm, and $\Delta = \pi/40$,

Fig. 2.5 Amplitudes of the induced moments, $|p_x|$, $|p_y|$, $|m_z|$, of the SRR as functions of ω/ω_u for different probing fields as shown in the figure. (From Ref. [33])

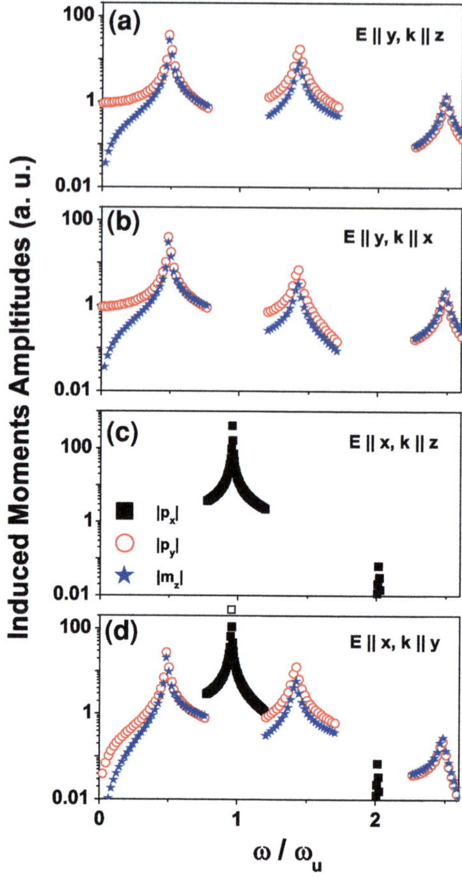

we first construct a 0.2 mm-thick metallic disk of radius 4.1 mm, then cut it by a 0.2 mm-thick air disk of radius 3.9 mm, and finally cut an air gap of the required width on the resulting structure. It is difficult to employ FDTD simulations to directly compute the dipole moments of a single SRR induced by an external plane wave. Instead, we study the transmission spectrum of an array of such SRR's, and identify the resonances by the dips of the transmission spectrum. For the two configurations studied in Fig. 2.5a, c, we construct SRR arrays to periodically tile the xy-plane, with lattice constants 16 mm along both x and y directions. For the other two configurations, we construct SRR arrays to periodically tile the xz- or yz- plane, respectively, with a lattice constant $= 12$ mm along x or y direction and 16 mm along z direction. The transmission spectra under the four plane wave inputs are shown, respectively in Fig. 2.6 a–d. When we compare Fig. 2.6 with the results shown in Fig. 2.5, we find that they agree with each other quite well. We clearly identify the dips at $\omega \approx 0.42\omega_0$ shown in Fig. 2.6a–d as the lowest eigen resonance mode (ω_1), the dips at $\omega \approx 1.19\omega_0$ shown in Fig. 2.6c, d as the second resonance mode (ω_2), and the dips at

Fig. 2.6 FDTD calculated transmission spectra of the SRR arrays as functions of ω/ω_u for plane wave inputs specified in the figure. Here the SRR has $R = 4$ mm, $a = 0.1$ mm, and $\Delta = \pi/40$. (From Ref. [32])

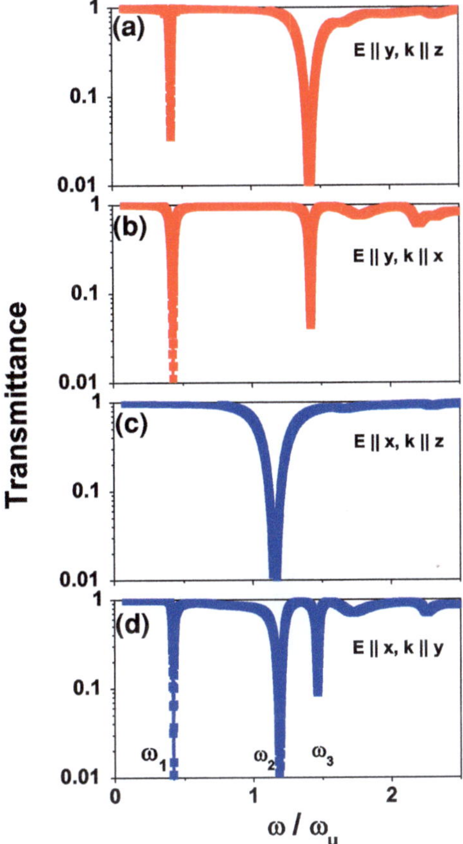

$\omega \approx 1.46\omega_0$ as the third one (ω_3). The quantitative differences between analytical and FDTD results must be caused by the couplings among different SRRs, since we have to adopt periodic arrays of SRRs to study the transmission spectra in the FDTD simulations.

2.7 Summary

In this chapter, we have established a rigorous mode-expansion theory for metallic systems in ring geometry in which the inductive/capacitive effects were included completely and the relevant circuit parameters were calculated rigorously. We have applied the theory to study the EM resonance properties of a single ring SRR. We found that the circuit problem can be analytically solved for ideal structures, leading to several useful analytical formulas for both resonance frequencies and current distributions. For general nonideal structures, we numerically solved the circuit equation and found that the obtained results match well with FDTD simulations on realistic structures. While there have been lots of efforts to study the SRRs theoretically, we believe an analytical theory established on more rigorous grounds is highly desirable, particularly since it can yield more physical insight and analytical formulas helpful for researchers working in this area.

Appendix

We describe the evaluation of the circuit elements in this Appendix. The current is along the ring, in component form, it is $j = (-\sin\varphi, \cos\varphi, 0)|j|$. We expand the current in a Fourier series as $|j| = \sum_m j_m \exp(im\varphi)$. We get $j_- = j_x - ij_y = \sum_m j_m \exp[i(m+1)\varphi]$.

In the QSA, the Green's function is proportional to $1/|r - r'| = 4\pi/(2l+1) \sum r_<^l Y_{lm}(\Omega) Y_{lm}^*(\Omega')/r_>^{l+1}$. We take the coordinate system so that the z axis is perpendicular to the ring. The ring is thus in the plane with $\theta=0$. Assume a time dependence of the form $\exp(i\omega t)$. For the self inductance of a single ring, the expansion parameter is $\alpha = r_</r_> = (R-a)/R$ where a is the radius of the wire that makes up the ring. For the mutual inductance of two concentric rings, the expansion parameter is $r_</r_> = R_1/R_2$. Carrying out the integration for $E_L^m = -i\omega \sum_m L_{mm'} I_{m'}$ $L_{mm'} = \frac{1}{2\pi} \int (\hat{e}_\phi \cdot \hat{e}_{\phi'}) e^{i(m'\phi'-m\phi)} g(\phi,\phi') \, R d\phi' d\phi/c^2$ we get, using the definition of the spherical harmonics $Y_{lm}=[(2l+1)(l-m)!/4\pi(l+m)!]^{1/2} P_l^m (\cos\theta)e^{im\phi}$,

$$E_{m-1} = -i\omega/c^2 \sum_l [P_l^m(0)]^2 [(l-m)!/(l+m)!] j_{m-1}\alpha^l/R \tag{A.1}$$

The inductance is thus given by

$$L_{m-1} = \sum_{l=m}^{\infty} A_l(m)\alpha^l/(Rc^2) \tag{A.1a}$$

where

$$A_l(m) = [P_l^m(0)]^2[(l-m)!/(l+m)!]. \tag{A.2}$$

We show below that for large l,

$$A_l(m) \approx \pi^{-1} l^{-1} 2. \tag{A.3}$$

This slow decrease of the coefficients $A_l(m)$ of the infinite series with respect to l leads to a log dependence on 1-α. We thus get

$$L_{m-1} = [B_{m-1} + \log(R/2b)/\pi]/(Rc^2) \tag{A.4}$$

with a nonlog dependent coefficient B_m.

Asymptotic Behavior of $\mathbf{A_l(m)}$

We now examine the asymptotic behavior of the coefficient of the infinite series. Now the associated Legendre polynomial at the origin is given by [68] $P_l^m(0) = 2^m \pi^{-1/2} \cos[\pi(l+m)/2]\Gamma((l+m+1)/2)/\Gamma[(l-m)/2+1]$. The asymptotic behavior of the Gamma function for large n is given by $\Gamma(n) \propto \exp(-n)n^{n-1/2}$. We get $P_l^m(0) = 2^m \pi^{-1/2} \cos[\pi(l+m)/2]R$ where the ratio of the Gamma functions is given by $R = \exp(-m+1/2)[(l+m+1)/2]^{(l+m+1)/2}/[(l-m)/2+1]^{(l-m)/2+1}$. R can be written as $R = (l/2)^{m-1/2}\exp[-m+1/2+[(l+m+1)\ln(1+(m+1)/l)/2) - (l-m+2)\ln[1+(2-m)/l]]/2$. In the large l limit we get $P_l^m(0) \approx \pi^{-1/2} \cos[\pi(l+m)/2]l^{m-1/2}2^{1/2}\exp[-m+1/2+(m+1)/2)-(2-m)/2]$. This can be simplified as

$$P_l^m(0) \approx \pi^{-1/2} \cos[\pi(l+m)/2]l^{m-1/2}2^{1/2}. \tag{A.5}$$

We thus have $A_m(l) \approx [P_l^m(0)]^2 \exp[2m + (l-m+1/2)\ln(1+(1-m)/l) - (l+m+1/2)\ln(1+(1+m)/l)]l^{-2m}$. In the large l limit, this becomes

$$A_m(l) \approx [P_l^m(0)]^2 \exp[2m + (1-m)) - (1+m))]l^{-2m}. \tag{A.6}$$

Substituting in Eq. (A.5), we get Eq. (A.3).

Numerical Evaluation

The inductances can be calculated recursively. $P_l^m(0) = 2^m \pi^{-1/2} \cos[\pi(l+m)/2]\Gamma((l+m+1)/2)/\Gamma[(l-m)/2+1]$. $P_m^m(0) = 2^m \pi^{-1/2} \cos(\pi m)\Gamma(m+1/2)/\Gamma(1) = 2^m \pi^{-1/2}(-1)^m[(1)(3)\ldots(2m-1)]\pi^{1/2}/2^m$. $A_m(m) = P_m^m(0)^2/(2m)! = [(1)(3)\ldots(2m-1)]^2/(2m)!$.

$$A_m(m) = (1)(3)\ldots(2m-1)/[(2m)(2m-2)\ldots 2].$$

$$P_{l+2}^m(0) = -P_l^m(0)(l+m+1)/(l-m+2).$$

$$A_{l+2}(m) = A_l(m)(l+m+1)^2[(l+2-m)(l+1-m)]/[(l+2+m)(l+1+m)](l-m+2)^2.$$

$$A_{l+2}(m) = A_l(m)[(l+m+1)(l+1-m)/[(l+2+m)(l-m+2)]].$$

Beyond QSA

It is possible to include the effect of the radiation resistance by using the formula $\exp(ik|r - r'|)/|r - r'| = 4\pi ik \sum_l j_l(kr_<)h_l(kr_>) \sum_m Y_{lm}(\Omega)Y_{lm}^*(\Omega')$.
Instead of Eq. (2.1a) we now have

$$L_{m-1} = \sum_{l=m}^{\infty} A_l(m)j_l(k|R - a|)h_l(kR)/(Rc^2) \qquad (A.7)$$

The formula for $A_l(m)$, Eq. (A.2) and its asymptotic form, Eq. (A.3) remains the same.

The leading correction to the QSA result is a damping term proportional to kR. This can also be seen trivially from the formula $\exp(ik|r - r'|)/|r - r'| \approx 1/|r - r'| + ik$.

This calculation can be easily generalized to the mutual inductance of two separate but concentric rings.

References

1. V.G. Veselago, Sov. Phys. Usp. **10**, 509 (1968)
2. L. Landau, E.M. Lifschitz, *Electrodynamics of Continuous Media* (Elsevier, New York, 1984)
3. J.B. Pendry, A.J. Holden, D.J. Robbins, W.J. Stewart, IEEE Trans. Microwave Theory Tech. **47**, 2075 (1999)
4. D.R. Smith, W.J. Padilla, D.C. Vier, S.C. Nemat-Nasser, S. Schultz, Phys. Rev. Lett. **84**, 4184 (2000)
5. R.A. Shelby, D.R. Smith, S. Schultz, Science **292**, 77 (2001)
6. H.O. Moser, B.D.F. Casse, O. Wilhelmi, B.T. Saw, Phys. Rev. Lett. **94**, 063901 (2005)
7. G. Dolling, M. Wegener, C.M. Soukoulis, S. Linden, Opt. Lett. **32**, 53 (2007)
8. G. Dolling, C. Enkrich, M. Wegener, C.M. Soukoulis, S. Linden, Opt. Lett. **31**, 1800 (2006)
9. S. Zhang, W. Fan, N.C. Panoiu, K.J. Malloy, R.M. Osgood, S.R.J. Brueck, Phys. Rev. Lett. **95**, 137404 (2005)
10. V.M. Shalaev, W. Cai, U.K. Chettiar, H. Yuan, A.K. Sarychev, V.P. Drachev, A.V. Kildishev, Opt. Lett. **30**, 3356 (2005)
11. J.B. Pendry, Phys. Rev. Lett. **85**, 3966 (2000)
12. N. Fang, H. Lee, C. Sun, X. Zhang, Science **308**, 534 (2005)
13. A. Ono, J. Kato, S. Kawata, Phys. Rev. Lett. **95**, 267407 (2005)
14. P.A. Belov, Y. Hao, S. Sudhakaran, Phys. Rev. B **73**, 033108 (2006)
15. N. Engheta, IEEE Antennas Wireless Propag. Lett. **1**, 10 (2002)
16. L. Zhou, H. Q. Li, Y. Q. Qin, Z. Y. Wei, and C. T. Chan, Appl. Phys. Lett. 86 101101 (2005)
17. H. Li, J. Hao, L. Zhou, Z. Wei, L. Gong, H. Chen, C. T. Chan, Appl. Phys. Lett. **89**, 104101 (2006)
18. I. Gil, J. Garcia–Garcia, J. Bonache, F. Martin, M. Sorolla, R. Marques, Electron. Lett. **40**, 1347 (2004)
19. K. Aydin1, I. Bulu1, K. Guven1, M. Kafesaki, C. M Soukoulis, and E. Ozbay, New J. Phys. **7**, 168 (2005).
20. A. Radkovskaya, M. Shamonin, C.J. Stevens, G. Faulkner, D.J. Edwards, E. Shamonina, L. Solymar, Microwave Opt. Technol. Lett. **46**, 473 (2005)

21. K.A. Boulais, D.W. Rule, S. Simmons, F. Santiago, V. Gehman, K. Long, A. Rayms-Keller, Appl. Phys. Lett. **93**, 043518 (2008)
22. K. Aydina, E. Ozbay, J. Appl. Phys. **101**, 024911 (2007)
23. K. Aydin, K. Guven, N. Katsarakis, C.M. Soukoulis, E. Ozbay, Opt. Express **12**, 5896 (2004)
24. B. Sauviac, C.R. Simovski, S.A. Tretyakov, Electromagnetics **24**, 317 (2004)
25. M. Shamonin, E. Shamonina, V. Kalinin, L. Solymar, J. Appl. Phys. **95**, 3778 (2004)
26. R. Marqués, F. Mesa, J. Martel, F. Medina, IEEE Trans. Antennas Propag. **51**, 2572 (2003)
27. F. Aznar, J. Bonache, F. Martin, Appl. Phys. Lett. **92**, 043512 (2008)
28. V. Zhurbenko, T. Jensen, V. Krozer, P. Meincke, Microwave Opt. Technol. Lett. **50**, 511 (2008)
29. M. Shamonin, E. Shamonina, V. Kalinin, L. Solymar, Microwave Opt. Technol. Lett. **44**, 133 (2005)
30. J. García-García, F. Martín, J.D. Baena, R. Marqués, L. Jelinek, J. Appl. Phys. **98**, 033103 (2005)
31. J.D. Baena, R. Marques, F. Medina, J. Martel, Phys. Rev. B **69**, 014402 (2004)
32. R. Marques, F. Medina, R. Rafii-El-Idrissi, Phys. Rev. B **65**, 144440 (2002)
33. L. Zhou, S.T. Chui, Phys. Rev. B **74**, 035419 (2006)
34. L. Zhou, S.T. Chui, Appl. Phys. Lett. **90**, 041903 (2007)
35. X.Q. Huang, Y. Zhang, S.T. Chui, L. Zhou, Phys. Rev. B **77**, 235105 (2008)
36. S.T. Chui, Y. Zhang, L. Zhou, J. Appl. Phys. **104**, 034305 (2008)
37. N. Katsarakis, T. Koschny, M. Kafesaki, E.N. Economou, C.M. Soukoulis, Appl. Phys. Lett. **84**, 2943 (2004)
38. P. Markos, C.M. Soukoulis, Phys. Rev. E **65**, 036622 (2002)
39. J. Zhou, Th Koschny, M. Kafesaki, E.N. Economou, J.B. Pendry, C.M. Soukoulis, Phys. Rev. Lett. **95**, 223902 (2005)
40. T.J. Yen, W.J. Padilla, N. Fang, D.C. Vier, D.R. Smith, J.B. Pendry, D.N. Basov, X. Zhang, Science **303**, 1494 (2004)
41. S. Linden, C. Enkrich, M.n Wegener, J. Zhou, T. Koschny, C. M. Soukoulis, Science **306**, 1351 (2004)
42. J. Zhou, T. Koschny, C.M. Soukoulis, Opt. Express **15**, 17881 (2007)
43. M.W. Klein, C. Enkrich, M. Wegener, C.M. Soukoulis, S. Linden, Opt. Lett. **31**, 1259 (2006)
44. E.N. Economou, Th Koschny, C.M. Soukoulis, Phys. Rev. B **77**, 092401 (2008)
45. Q. Zhao, L. Kang, B. Du, B. Li, J. Zhou, H. Tang, X. Liang, B. Zhang, Appl. Phys. Lett. **90**, 011112 (2007)
46. C. Enkrich, M. Wegener, S. Linden, S. Burger, L. Zschiedrich, F. Schmidt, J.F. Zhou, Th Koschny, C.M. Soukoulis, Phys. Rev. Lett. **95**, 203901 (2005)
47. F. Bilotti, A. Toscano, L. Vegni, IEEE Trans. Antennas Propag. **55**, 2258 (2007)
48. A.K. Azad, A.J. Taylor, E. Smirnova, J.F. O'Hara, Appl. Phys. Lett. **92**, 011119 (2008)
49. D.R. Smith, D.C. Vier, N. Kroll, S. Schultz, Appl. Phys. Lett. **77**, 2246 (2000)
50. D.R. Smith, S. Schultz, P. Markos, C.M. Soukoulis, Phys. Rev. B **65**, 195104 (2002)
51. B. Popa, S.A. Cummer, Phys. Rev. B **72**, 165102 (2005)
52. K.B. Alicia, E. Ozbay, J. Appl. Phys. **101**, 083104 (2007)
53. B.E. Little, S.T. Chu, H.A. Haus, J. Foresi, J.-P. Laine, J. Lightwave Technol. **15**, 998 (1997)
54. R. Marques, J. Martel, F. Mesa, F. Medina, Phys. Rev. Lett. **89**, 183901 (2002)
55. H. Xu, Z. Wang, J. Hao, J. Dai, L. Ran, J.A. Kong, L. Zhou, Appl. Phys. Lett. **92**, 041122 (2008)
56. M.C.K. Wiltshire, J.B. Pendry, I.R. Young, D.J. Larkman, D.J. Gilderdale, J.V. Hajnal, Science **291**, 849 (2001)
57. M.C.K. Wiltshire, J.V. Hajnal, J.B. Pendry, D.J. Edwards, C.J. Stevens, Opt. Express **11**, 709 (2003)
58. E. Verney, B. Sauviaca, C.R. Simovski, Phys. Lett. A **331**, 244 (2004)
59. J.D. Baena, L. Jelinek, R. Marqués, J. Zehentner, Appl. Phys. Lett. **88**, 134108 (2006)
60. J.D. Baena, L. Jelinek, R. Marqués, J.J. Mock, J. Gollub, D.R. Smith, Appl. Phys. Lett. **91**, 191105 (2007)

61. J.D. Baena, L. Jelinek, R. Marqués, Phys. Rev. B **76**, 245115 (2007)
62. P. Gay-Balmaz, O.J.F. Martin, Appl. Phys. Lett. **81**, 939 (2002)
63. Q. Zhao, L. Kang, B. Du, H. Zhao, Q. Xie, X. Huang, B. Li, J. Zhou, L. Li, Phys. Rev. Lett. **101**, 027402 (2008)
64. Th Koschny, L. Zhang, C.M. Soukoulis, Phys. Rev. B **71**, 121103 (2005)
65. J.D. Jackson, *Classical Electromagnetic*, 3rd edn. (Wiley, New York, 1999)
66. CONCERTO 4.0, Vector Fields Limited, England, 2005
67. See, for example, Matthews and Walker, Mathematical Physics, Benjamin, New York, 1965, pp. 50–51, Eq. (2.33)
68. Handbook of Mathematical functions by Abramowitz and Stegun Eq. (8.6.1)

Chapter 3
Resonance Properties of Metallic Ring Systems: More Complex Structures

3.1 Introduction

When we developed the mode-expansion theory in the last chapter, we have assumed that the metal wire forming the ring possesses a circular cross-section and the system contains only one metallic ring. However, in reality, the metallic wires often exhibit flat rectangular cross-sections [1–6], and the SRRs fabricated or proposed often have two (or more) metallic rings, which are either lying on the same plane coupled horizontally (denoted as coplanar double-ring SRR) [1–4] or lying on different planes coupled vertically (denoted as broadside-coupled (BC) SRR) [5, 6]. Typically, the small gaps are opened at opposite positions on two rings in these double-ring systems, as shown in Fig. 3.1. Why did such double-ring systems draw intensive attention?

The SRRs were proposed as the magnetic building blocks of left-handed materials (LHM) [1, 2]. Since the properties of the LHM was arrived at based on the concept of effective medium theory, to make an LHM with good quality, one naturally requires the building block to be of a size as small as possible when compared with the wavelength. Furthermore, the interpretation is simpler if its *bi-anisotropy* is as small as possible. However, as we learned from Chap. 2, a single-ring SRR exhibits strong *bi-anisotropy*, and the largest resonance wavelength of a single-ring SRR is of the order of 2π times the diameter of the ring, not very large when compared with its size. The double-ring SRRs were proposed to overcome these two shortcomings [1–6], and thus were believed as better candidates as the magnetic building blocks for LHM.

In this chapter, we shall extend the mode-expansion theory developed in the last chapter to more complex and realistic situations, in which the system can possess more metallic rings and the wire cross-section can be of other shape such as a flat rectangle. We then apply the extended theory to study the rich resonance properties of a coplanar double-ring SRR (Sect. 3.2) [7] and a BC-SRR (Sect. 3.3) [8]. We conclude this chapter in Sect. 3.4.

S. T. Chui and L. Zhou, *Electromagnetic Behaviour of Metallic Wire Structures*, DOI: 10.1007/978-1-4471-4159-4_3, © Springer-Verlag London 2013

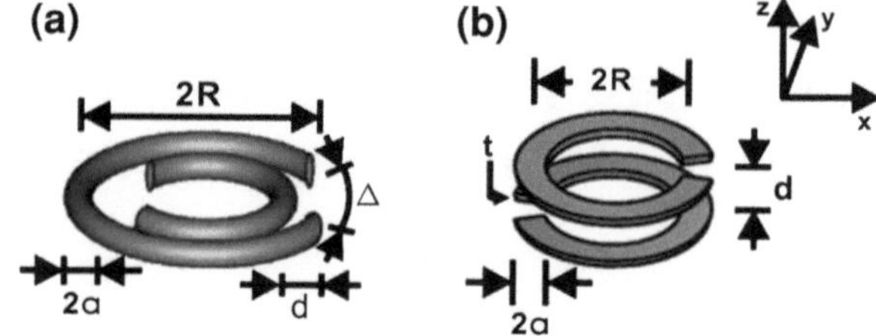

Fig. 3.1 Geometries of (a) coplanar SRR and (b) BC–SRR made with flat wires

3.2 Double-Ring SRR

The theory established in Chap. 2 can be easily extended to more complex situations, such as systems with more metallic rings. For the coplanar double-ring SRR as shown in Fig. 3.1a, we find that the original matrix-circuit Eq. (2.13) should be rewritten as

$$\sum_{m'j'} H_{\{mj\},\{m'j'\}} I_{m'j'} = E_{\text{ext}}^{mj} \tag{3.1}$$

where the matrix elements are

$$H_{\{mj\}.\{m'j'\}} = \rho_j(m - m')\delta_{jj'} + i\left(\omega L_m^{jj'} - \frac{1}{\omega C_m^{jj'}}\right)\delta_{mm'}, \tag{3.2}$$

and E_{ext}^{mj} denotes the projection of external field on the mth Fourier component of the jth metallic ring. $\rho_j(m - m')$ is the Fourier component of the resistivity function of the jth ring, which can be easily derived once the geometrical parameters are given. Thanks to the ring geometry, the inductance and capacitance matrixes are diagonal with respect to the index m. L_m^{jj} and C_m^{jj} are the self-inductive/capacitive parameters of the jth ring, which have been explicitly given in Eq. (2.11). $L_m^{jj'}$ and $C_m^{jj'}$, the mutual inductance/capacitance parameters between two rings, can be calculated similarly based on the theory developed in the last chapter. Explicitly, suppose the radius of the jth ring is R_j and each ring has the same wire radius a, we found that

$$L_m^{j,i} = \pi(A_{m-1}^{j,i} + A_{m+1}^{j,i})/c^2, \quad 1/C_m^{j,i} = 2m^2\pi A_m^{j,i}/(R_i R_j) \tag{3.2a}$$

where

$$A_m^{j,i} = \sum_{l=|m|} \frac{(l - m)!}{(l + m)!}(\tilde{\alpha})^l \left[P_l^m(0)\right]^2 \tag{3.2b}$$

with $\tilde{\alpha} = \min(R_i, R_j)/\max(R_i, R_j)$. It is easy to demonstrate that the inductance/capacitance matrixes are symmetrical ones, i.e., $L_m^{i,j} = L_m^{j,i}$, $C_m^{i,j} = C_m^{j,i}$. Therefore, similar to the single-ring case, one can solve the matrix equation (3.1) to obtain the resonance properties of a double-ring system. We first describe in this section the analytical solution of Eq. (3.1) under some limiting conditions, and then present our numerical investigations of Eq. (3.1) for two particular examples in the following two sections.

Under the same three conditions as mentioned in Sect. 2.5 of last chapter, we found that the circuit equation to determine the resonance frequencies is given by $\mathbf{HI} = (\mathbf{H_0}+\mathbf{Y})\mathbf{I} = 0$. The unperturbed and the perturbation matrices are

$$\mathbf{H_0} = \begin{pmatrix} r\mathbf{M^a} & \mathbf{0} \\ \mathbf{0} & r\mathbf{M^b} \end{pmatrix}, \quad \mathbf{Y} = \begin{pmatrix} \mathbf{X} & \mathbf{X'} \\ \mathbf{X'} & \mathbf{X} \end{pmatrix} \tag{3.3}$$

where the matrix $\mathbf{M^a}$ (with elements $\mathbf{M^a}_{m,m'} = 1$) is for ring 1 and the matrix $\mathbf{M^b}$ (with elements $\mathbf{M^b}_{m,m'} = (-1)^{m-m'}$) is for ring 3. $\mathbf{X}, \mathbf{X'}$ are diagonal matrices with X_m same as before and $X'_m = i[L'_m\omega - m^2/(C'_m\omega)]$, in which L'_m and C'_m are the mutual inductances and capacitances between two rings. Here, we have neglected the size differences of the two rings in writing the self-interaction terms. Again, there are even and odd modes under the transformation m to $-m$. Here, we focus only on the even modes. Following the perturbation theory, we first consider the unperturbed matrix problem $\mathbf{H_0}I^{(0)}= \mathbf{0}$. The solutions can be written as $I^{(0)}=\begin{bmatrix} I^{(0a)} \\ I^{(0b)} \end{bmatrix}$ which satisfy

$$\sum_m I^{(0a)}{}_m = 0, \sum_m (-1)^m I^{(0b)}{}_m = 0. \tag{3.3a}$$

Now assume the full solution as $\mathbf{I} = I^{(0)}+I^{(1)}$, where $I^{(1)}$ is of the order of $1/r$, we get $\mathbf{HI} = \mathbf{Y}I^{(0)}+\mathbf{H_0}I^{(1)} + o(r^{-1}) = 0$. Putting Eq. (3.3) into the above matrix, we find

$$\begin{cases} I^{(0a)}{}_m = \left[X_m\sigma^a - (-1)^m X'_m\sigma^b\right]/\left[X_m^2 - X_m'^2\right] \\ I^{(0b)}{}_m = \left[(-1)^m X_m\sigma^b - X'_m\sigma^a\right]/\left[X_m^2 - X_m'^2\right] \end{cases} \tag{3.3b}$$

where $\sigma^a = -r\sum_{m'} \varepsilon^a{}_{m'}, \sigma^b = -r\sum_{m'} (-1)^{m'} \varepsilon^b{}_{m'}$ are two constants (the localized electric fields) independent of index m. Applying the constrains (3.3a), we finally arrive at the equation

$$\sum_m \frac{1}{X_m \mp (-1)^m X'_m} = 0 \tag{3.4}$$

to determine the resonance frequencies of the double-ring system. We note that this equation recovers the single-ring results [i.e., Eq. (2.19a)] in the absence of mutual-interaction terms (i.e., $X'_m = 0$). In explicit form, this equation becomes

$$\sum_{m=1}^{\infty} \frac{2\omega^2(L_0 \pm L'_0) - 2ir_c\omega}{i\omega r_c - L_m\omega^2 + m^2/C_m \pm (-1)^m\left[-L'_m\omega^2 + m^2/C'_m\right]} = 1 \tag{3.4a}$$

Again, we emphasize that Eq. (3.4) is the **exact** solution of the double-ring problem (recalling $r \to \infty$).

The current can be obtained by inverting the circuit equation as

$$I_{1m} = [X_m(E_{\text{ext},1m} + \sigma_1) - (E_{\text{ext},2m} + (-1)^m \sigma_2)X'_m]/(X_m^2 - X_m'^2);$$
$$I_{2m} = [X'_m(E_{\text{ext},1m} + \sigma_1) - (E_{\text{ext},2m} + (-1)^m \sigma_2)X_m]/(X_m^2 - X_m'^2).$$

(3.4b)

The external fields $(E_{\text{ext},1}, E_{\text{ext},2})$ can be decomposed as a sum of fields of opposite "parities"

$$E_{\text{ext},\pm m} = (E_{\text{ext},1m} \pm (-1)^m E_{\text{ext},2m})/2$$

(3.4c)

so that $E_{\text{ext},1m} = E_{\text{ext},+m} + E_{\text{ext},-m}$, $E_{\text{ext},2m} = (-1)^m(E_{\text{ext},+m} - E_{\text{ext},-m})$. In this basis from the boundary conditions we get the localized electric fields

$$\sigma_{1\pm} = -[\sum_m E_{\text{ext},\pm m}/(X_m \pm (-1)^m X'_m)]/\sum_m 1/(X_m \pm (-1)^m X'_m).$$

(3.4d)

3.3 Coplanar Double-Ring SRR

In general cases where the analytical formulas are not strictly applicable, we can numerically solve the circuit Eq. (3.1) to obtain the resonance properties of such systems. Consider a coplanar double-ring SRR with ring radii $R_1 = R, R_2 = R - d$, and wire radii $a_1 = a_2 = a \ll R$. We first describe our numerical investigation of Eq. (3.1), then compare our results with analytical estimates using Eq. (3.4a), and finally compare with the FDTD simulation results. For an example with structural details shown in the caption of Fig. 3.2a, we solve numerically the matrix Eq. (3.1) and show in Fig. 3.2a the calculated $\min[|\lambda_l|]$ as a function of ω/ω_u. Compared with the spectra of a single-ring SRR shown in the same figure, we find that each single-ring mode has split into a pair of modes in the double-ring case through mutual inductance/capacitance effects. The mode pairs involve similar sets of Fourier components like their corresponding single-ring modes. We depict in Fig. 3.2b, c the eigenvectors for the first two modes, which are shown to involve mainly the $m = 0, \pm 1$ components, similar to the lowest single-ring mode. It is interesting to note that while the $m = 0$ components are in-phase (out of phase) for the $\omega_1^L (\omega_1^H)$ mode, the $m = \pm 1$ components behave in just the opposite manner! Noting from Eq. (2.31) that the $m = 0$ components contribute to the magnetic polarization (m_z) while the $m = \pm 1$ components contribute to the electric polarization (p_x, p_y), we immediately understand that the magnetic (electric) moments contributed by the two rings significantly cancel each other for the $\omega_1^H (\omega_1^L)$ mode. Therefore, the total magnetic (electric) polarizations are highly diminished in the $\omega_1^H (\omega_1^L)$ mode and the bi-anisotropy is also highly suppressed as the two rings approach each other.

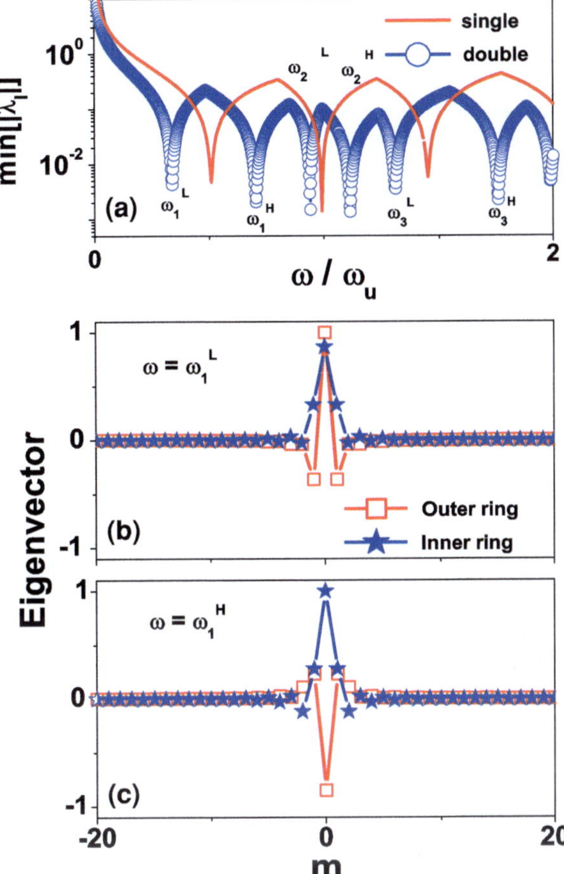

Fig. 3.2 a $\min[\|\lambda_l\|]$(in arbitrary units) as the functions of ω/ω_u calculated for a double-ring SRR (*symbols*) with $R = 4\,\mathrm{mm}$, $d = 0.4\,\mathrm{mm}$, $a = 0.1/\sqrt{\pi}\,\mathrm{mm}$, $\Delta = \pi/90$, and a single-ring SRR (*lines*) with $R = 3.8\,\mathrm{mm}$, $a = 0.1/\sqrt{\pi}\,\mathrm{mm}$, $\Delta = \pi/90$. Eigenvector component $Q_l^{i,m}$ as the functions of m for the outer ring ($i = 1$, *open squares*) and the inner ring ($i = 2$, *solid stars*) for the resonance mode at ω_1^L (**b**), and at ω_1^H (**c**). (Reprinted with permission from Ref. [7]. Copyright (2007), American Institute of Physics.)

Figure 3.3 shows the dipole moments of the structure induced by different external plane waves. Similar to the single-ring case (Fig. 2.5), we find that the odd-numbered modes $\left(\omega_1^L, \omega_1^H, \omega_3^L, \omega_3^H, \ldots\right)$ carry both electric (P_y) and magnetic (M_z) polarizations, while the even-numbered ones $\left(\omega_2^L, \omega_2^H, \omega_4^L, \omega_4^H, \ldots\right)$ carry only electric (P_x) polarizations. In particular, one may easily find that m_z is much stronger than p_y for the mode ω_1^L, but things become reversed for mode ω_1^H. These are precisely the evidences of the polarization diminishment effects discussed above. Similar to the single-ring SRR case, we have also performed FDTD simulations to calculate the transmission spectra through arrays of realistic double-ring SRR structures, and again the transmission dips in the obtained transmission spectra (Fig. 3.4) exhibit excellently one-by-one correspondence to the moment peaks in Fig. 3.3.

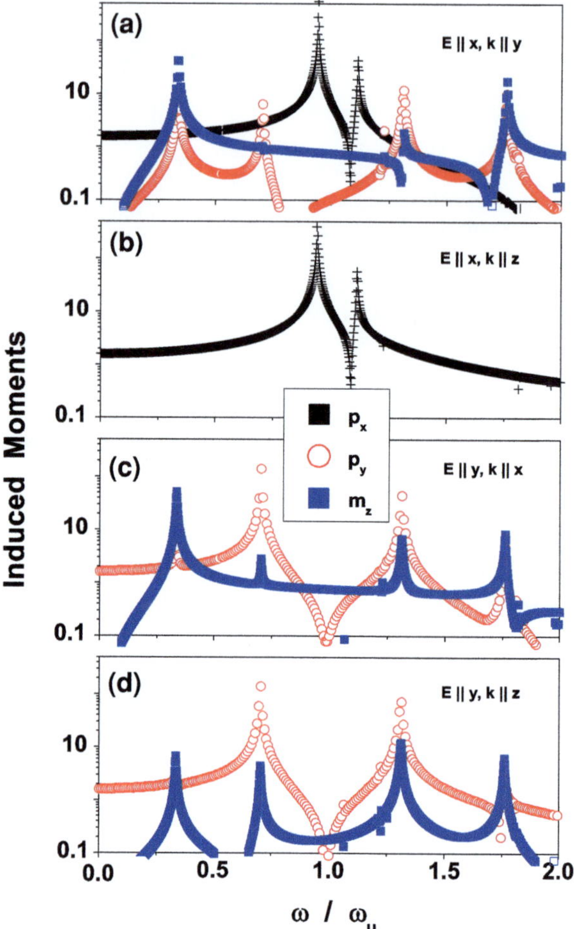

Fig. 3.3 Dipole moments (in arbitrary units) of the double-ring SRR (same as Fig. 3.2) induced by external plane waves with $\vec{E}$ and $\vec{k}$ directions specified in the legends. (Reprinted with permission from Ref. [7]. Copyright (2007), American Institute of Physics.)

The present theory allows us to *quantitatively* examine the bi-anisotropy of the SRR. Based on the definition in Refs. [5, 6], we find that the constitutional relation of a double-ring SRR can be generally written as

$$\begin{cases} p_x = \alpha_{xx}^{ee} E_x^{\text{ext}} \\ p_y = \alpha_{yy}^{ee} E_y^{\text{ext}} + i\alpha_{yz}^{em} B_z^{\text{ext}} \\ m_z = -i\alpha_{yz}^{em} E_y^{\text{ext}} + \alpha_{zz}^{mm} B_z^{\text{ext}} \end{cases} \tag{3.5}$$

where $\alpha_{xx}^{ee}, \alpha_{yy}^{ee}$ are the electric polarizabilities, α_{zz}^{mm} is the magnetic polarizability, and α_{yz}^{em} is the bi-anisotropic polarizability. We calculated the induced dipole moments of the SRR under the following four types of excitations, which are plane waves with (a) $\vec{E} \| \hat{y}, \vec{k} \| \hat{z}$, (b) $\vec{E} \| \hat{y}, \vec{k} \| \hat{x}$, (c) $\vec{E} \| \hat{x}, \vec{k} \| \hat{z}$, (d) $\vec{E} \| \hat{x}, \vec{k} \| \hat{y}$, and then retrieved the parameters $\alpha_{xx}^{ee}, \alpha_{yy}^{ee}, \alpha_{zz}^{mm}, \alpha_{yz}^{em}$, based on Eq. (3.5). We show in Fig. 3.5a, b the

Fig. 3.4 FDTD-simulated transmission spectra of double-ring SRR arrays under plane wave inputs with $\vec{E}$ and $\vec{k}$ directions specified in the legends. {Reprinted with permission from Ref. [7]. Copyright (2007), American Institute of Physics.}

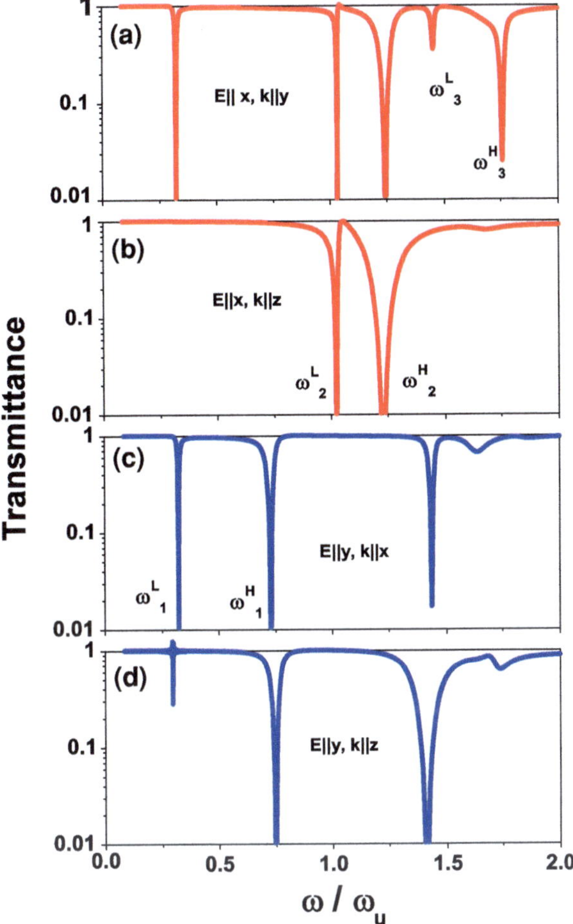

calculated values of electric polarizability $\left(\alpha^{ee}_{yy}\right)$ and bi-anisotropic polarizability $\left(\alpha^{em}_{yz}\right)$, respectively, for double-ring SRR's with different values of d. Clearly, both α^{ee}_{yy} and α^{em}_{yz} are significantly suppressed for the fundamental mode as d decreases. As a result, such a mode becomes *purely magnetic* in the limit of $d \to 0$.

The magnetic resonance frequency ω^{L}_{1} is drastically lowered compared to the single-ring mode (denoted by ω_{1}). Solid symbols in Fig. 3.6a are the calculated relative frequency shift, i.e., $\delta\omega/\omega_{1} = (\omega^{L}_{1} - \omega_{1})/\omega_{1}$, as a function of d for two sets of samples. As expected, $\delta\omega/\omega_{1}$ becomes larger as d decreases. FDTD simulations are performed on a series of samples with $a/R = 0.1/(4\sqrt{\pi})$, and the results are depicted in Fig. 3.7a as open stars. We find excellent agreement between the FDTD and theoretical calculations for $m = 0, \pm 1 \delta\omega/\omega_{1}$.

Fig. 3.5 Amplitudes of (**a**) electric polarizability $|\alpha_{yy}^{ee}|$ and (**b**) bi-anisotropic polarizability $|\alpha_{yz}^{em}|$ as the functions of ω/ω_u, for double-ring SRR's with $R = 4$mm, $a = 0.1/\sqrt{\pi}$mm, and different values of d specified in the legend. (Reprinted with permission from Ref. [7]. Copyright (2007), American Institute of Physics.)

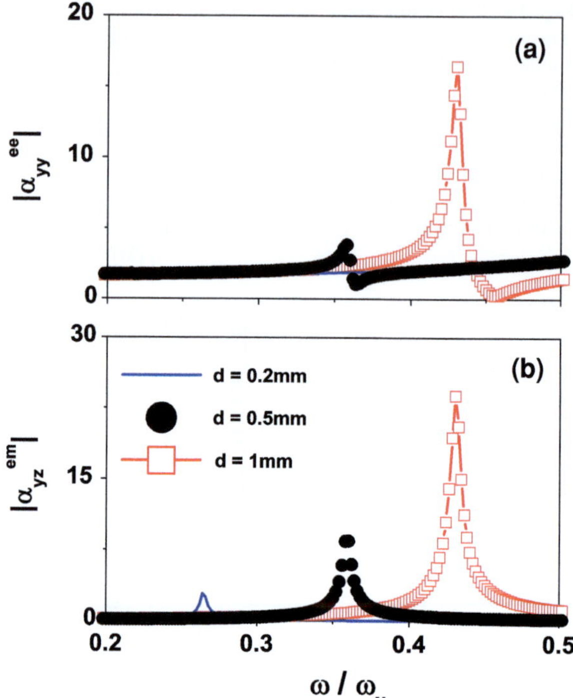

We next derive an analytic approximate expression for $\delta\omega m/\omega_1$. As Fig. 3.2b, c indicates that the ω_1^L and ω_1^H modes mainly involve the $m = 0, \pm 1$ Fourier components, we thus retain only those terms in the H-matrix defined in Eq. (2.1), since the contributions of higher order terms decrease significantly as m increases. When d is small enough, it is safe to set $\Omega_m^{11} \approx \Omega_m^{22} = \Omega_m, L_m^{11} \approx L_m^{22} = L_m$. From Eqs. (3.4) and (3.4a) we get

$$\omega_1^\pm = \sqrt{(L_1\Omega_1^2 \pm \tilde{L}_1\tilde{\Omega}_1^2)/(2L_0 + L_1 \mp (2\tilde{L}_0 - \tilde{L}_1))} \qquad (3.6)$$

for the lowest two resonance frequencies. Here, we set $\tilde{\Omega}_m = \Omega_m^{12} = \Omega_m^{21}, \tilde{L}_m = L_m^{12} = L_m^{21}$. The fundamental magnetic resonance frequency is $\omega_1^L = \omega_1^-$. Under the conditions $a \ll R, d \ll R$, we find that $L_m \approx -2\ln(2a/R)$, $\tilde{L}_m \approx -2\ln(2d/R)$, and $\Omega_1 \approx \tilde{\Omega}_1 \approx \omega_0$, according to the theory presented in Chap. 2. In the limit of $R > d \gg a$, we can expand $\omega_1^\pm$ in Eq. (3.6) to a series of $\tilde{L}_m/L_{m'}$. Keeping the lowest term, we obtain a formula,

$$\omega_1^L = \omega_1[1 - 2\ln(2d/R)/(3\ln(2a/R))], \qquad (3.7)$$

to determine the resonance frequency in *explicit* terms of all geometrical parameters of an SRR (a, d, R, etc.). The physics behind Eq. (3.7) is that $\delta\omega/\omega_1$ is

Fig. 3.6 a $\delta\omega/\omega_1$ as a function of d/R for two sets of double-ring systems with different values of a/R, calculated by the mode-expansion theory (*solid symbols*), FDTD simulations (*open symbols*), and Eq. (3.7) (*lines*) **(b)** The limiting frequency achieved in a double-ring SRR as the functions of a/R, calculated by the mode-expansion theory (symbols) and the analytical formula Eq. (3.8). (Reprinted with permission from Ref. [7]. Copyright (2007), American Institute of Physics.)

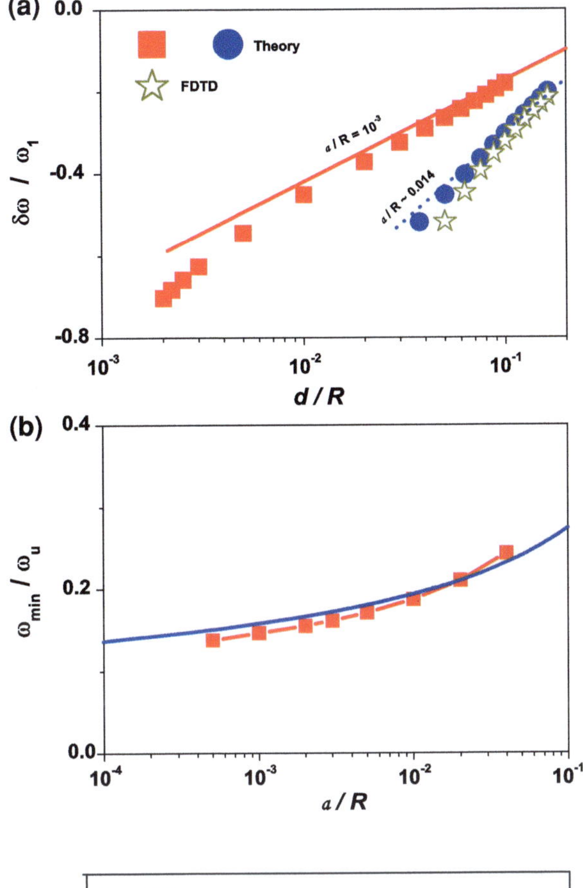

Fig. 3.7 SRR resonance frequency f (measured in GHz) as a function of t (setting $w = 0.9$mm, *solid line*) and w (setting $t = 0.2$mm, *dotted line*), calculated by our analytical formula. Symbols are experimental data taken from Figs. 7 and 8 of Ref. [3]. (Reprinted with permission from Ref. [7]. Copyright (2007), American Institute of Physics.)

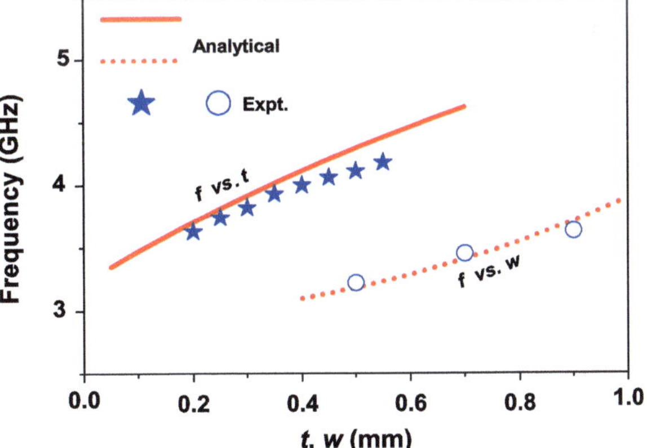

dictated by a competition between the mutual interaction and the self-interaction ($\propto \ln(2a/R)$). The lines in Fig. 3.6a are results calculated with such a formula, which agree well with both theoretical and FDTD results, particularly when $d \gg a$. Deviations exist as d is comparable to a, where a perturbation treatment fails.

Equation (3.6) indicates that $\omega_1^L \propto \sqrt{L_1 \Omega_1^2 - \tilde{L}_1 \tilde{\Omega}_1^2}$, thus ω_1^L can be significantly lowered when $d \to a$. We can derive an analytical formula to describe the fundamental resonance frequency in such a limit. For a system with fixed R and a, the minimum value of d is $2a$. Putting $d = 2a$ directly into Eq. (3.6), we get the following formula,

$$\omega_{\min} \xrightarrow{d \to 2a} \omega_0 \sqrt{\ln(1/2)/[4\ln(a/R)]}, \tag{3.8}$$

to estimate the smallest resonance frequency of such a structure. As $a/R \to 0$, this frequency approaches zero, indicating that the double-ring SRR can be sufficiently *subwavelength* in this limit! Figure 3.6b shows that $\omega_{\min}$ calculated based on the complete theory can indeed be well described by the formula (3.8).

Our formulae can be directly compared with experiments. For the SRR's studied experimentally in Ref. [3], adopting their definitions (i.e., set $R = r - 0.5w$, $a = w, d = t + w$ where w is the metal line width and t is the gap distance between two rings [3]) and taking the experimental data $\omega_1 = 2\pi \cdot 4.58 \text{GHz}$ from [3], we get an expression,

$$f = 4.58[1 - 2\ln((t + w)/(r - 0.5w))/[3\ln(w/(r - 0.5w))]] \tag{3.9}$$

for the resonance frequencies of the SRRs studied in Ref. [3]. As shown in Fig. 3.7, *without any adjustable parameters*, the experimental data taken from Ref. [3] are reasonably described by the analytical formula Eq. (3.9). It is worth mentioning that the SRR fabricated in Ref. [3] did not exhibit circular wire cross-sections. However, the good agreement displayed in Fig. 3.7 indicates that the analytical formula does capture the core physics of the frequency splitting, which is rather independent on the structure details.

3.4 BC–SRR

We now extend the mode-expansion theory to study the BC–SRR as depicted in Fig. 3.1b, in which the metallic wires possess flat rectangular cross-sections defined by $t \times 2a$ where $2a$ and t are the width and thickness of the wire. The distance between rings in the same plane has to be larger than width. No such constraint exists for rings not in the same plane but vertically on top of each other. Thus, it is easier to create deep subwavelength structure with this geometry. This structure also exhibits a magnetic moment in the plane of the ring, which can be useful for some applications. The circuit equation for such a system is still Eq. (3.1), but the circuit parameters need to be re-examined.

Obviously, the delta-like current distribution (Eq. (2.1)) may no longer be a reasonable assumption for flat wires, and we need to consider more realistic forms of the current distributions. In general, the current distributions over the rectangular cross-section should be very complicated and depend on the concrete form of the probing field. We consider a case so that the current is localized as two current sheets at the two surfaces. This happens when the thickness is larger than the skin depth: $t \gg \delta = 3.76\lambda^{1/2}\mu m$[See Eq. (2.10)] at a wavelength of λm Since $t \ll 2a$, we can neglect the current flowing along the two side surfaces located at $r = R \pm a$. In this section we describe an example with $t \gg \delta$ and model the true current distributions as two identical current sheets located at $z = \pm t/2$. It is straightforward to extend this calculation to the case $t < \delta$ for a single current sheet. Noting that $a \ll \lambda$, we further take an approximation to assume that the current distributes *uniformly* along the wire width. Collecting all these considerations, we can *approximate* the true current distribution in a flat wire as:

$$\vec{j}(\vec{r}) = \begin{cases} j(\phi) \cdot t_\delta[\delta(z+t/2) + \delta(z-t/2)]\vec{e}_\phi, & \rho \in [R-a, R+a] \\ 0, & \text{otherwise} \end{cases} \quad (3.10)$$

where $t_\delta \ll t$ is the skin depth of the metal. Based on the approximate form of current distribution (3.10) in flat wires, we appropriately average the **E** fields as well as the current $\vec{j}(\vec{r})$over the cross-section, $\bar{E}_{L,C}(\phi) \propto \int \vec{E}_{L,C}(\vec{r}) \cdot \vec{e}_\phi dS', I(\phi) \propto \int \vec{j}(\vec{r}) \cdot \vec{e}_\phi dS'$, and then derive the circuit parameters based on Fourier transformation and the techniques developed in this chapter. For such flat wires, we found a set of modified circuit parameters (L_m, C_m, etc.) which are different from those for the wire with circular cross-section. The derivations are rather complicated and we recommend the interested readers to refer to the original publication [8] for the explicit forms of the circuit parameters and the detailed mathematical derivations. With the modified circuit parameters Eq. (3.4) and its approximation, Eq. (3.6), are still applicable to the present case. We now apply our theory to study the EM resonance properties of the BC–SRR as shown in Fig. 3.1b.

From Eq. (3.6), the reduction in the resonance frequency comes from a cancelation between the mutual and the self-capacitances. For rings made with flat wires not on the same plane, the separation d between rings needs to be larger than the thickness t of each ring. In general t can be much smaller than width $2a$ of the flat wires, so that the difference of capacitances, and in turn, the resonance frequency, can be much smaller.

This conclusion remains valid when higher modes are included. To illustrate, when the $m = \pm 2$ modes are included, the resonance frequency is given by, for $r_c = 0$, $\omega_0'^2 = (1-\delta)(1/C_1 - 1/C_1')/[2(L_0 + L_0') + (1-\delta)(L_1 - L_1')]$ where $\delta = 2\omega_0^2(L_0 + L_0')/(1/C_2 + 1/C_2')$. Thus, as $1/C_1 - 1/C_1'$ is made small, the resonance frequency becomes small as well.

With the circuit parameter for the flat wire case, when both a and R are fixed, the normalized inverse-capacitance differences,$1 - C_m/C_m'$, approach zero as $d \to 0$, as shown in Fig. 3.8a calculated for $a/R = 0.025$. As $d \to 0$, $\sin\beta \to 0, r \to \rho, r' \to \rho'$; thus $1/C_m' \to 1/C_m$. As a result, the lowest resonance frequency (ω_0) of the double-ring SRR is significantly reduced as $d \to 0$, as shown in Fig. 3.8b by

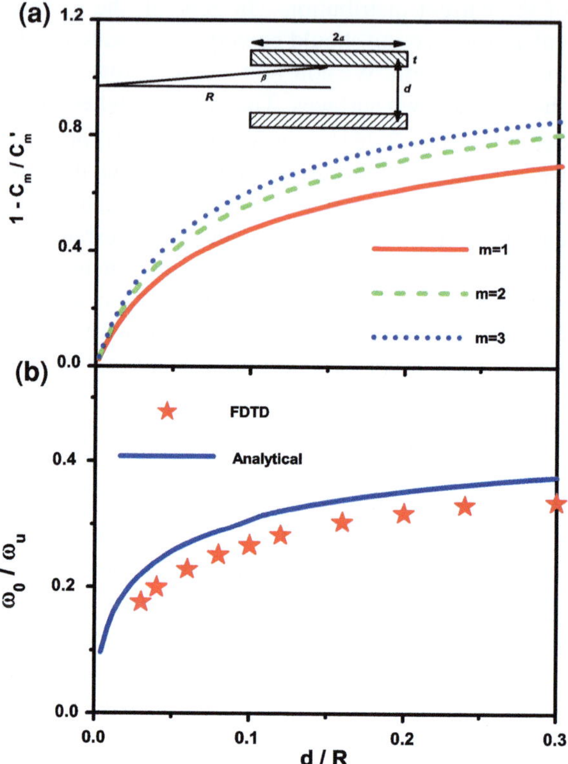

Fig. 3.8 a Differences between the inverses of the self- and mutual capacitances as a function of the ring–ring distance, for a double-ring SRR. The inset schematically shows the geometry of a double-ring SRR, where the two shaded areas represent the cross-sections of the two flat wires. **b** The lowest resonance frequency of the double-ring SRR as a function of the ring–ring distance, calculated by the present analytical theory (line) and the FDTD simulations on realistic structures. (Reprinted with permission from Ref. [8]. Copyright (2008), American Institute of Physics.)

the solid line, obtained by numerically solving Eq. (3.4) setting $a/R = 0.025$ with the circuit parameters described above and similar expressions for L_m and L'_m.

FDTD simulations have been performed on a series of realistic double-ring SRR's made with flat wires with different separation d. The FDTD-calculated ω_0 are shown in Fig. 3.8b as solid stars which agree quite well with the analytic solutions. The small discrepancies between the FDTD and the analytic results can be attributed to the approximations adopted in calculating the circuit parameters.

However, FDTD simulations are difficult to perform for the very small d cases, since in such cases the basic mesh discretizing the structure becomes too fine. Fortunately, analytic formulas are available when $d \to 0$. We find that $1 - C_1/C'_1 \approx F \cdot (d/R)^2$, where F is a dimensionless coefficient depending only weakly on a/R. Considering the three-mode expression as shown in Eq. (3.6), we further find that $\omega_0/\omega_u \approx \tilde{F} \cdot (d/R)$ with $\tilde{F}$ being another dimensionless parameter. Shown in Fig. 3.9 are $1 - C_1/C'_1$ and ω_0/ω_u as functions of d/R, calculated with the full theory setting $a/R = 0.025$. These numerical results accurately confirmed the above two formulas, and suggested that $\tilde{F} = 66$ and $F = 40240$ for the present structure.

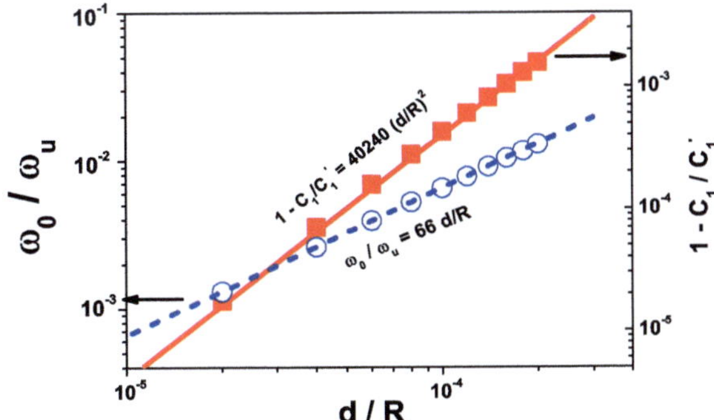

Fig. 3.9 ω_0/ω_u and $1 - C_1/C_1'$ as the functions of d/R in the limit of very small d/R, calculated based on the analytical theory (*symbols*) for $a/R = 0.025$, and the fitting formulas shown in the figure. (Reprinted with permission from Ref. [8]. Copyright (2008), American Institute of Physics.)

Table 3.1 Characteristics of electric/magnetic polarizations of low-lying modes for a BC–SRR and a single-ring SRR

System	Mode	p_x	p_y	m_z	m_x	m_y
Single	ω_1	No	Yes	Yes	No	No
Double	ω_1^L	No	Disappear	Enhanced	*New*	No
	ω_1^H	No	Enhanced	Disappear	No	No
Single	ω_2	Yes	No	No	No	No
Double	ω_2^L	Enhanced	No	No	No	No
	ω_2^H	Disappear	No	No	No	*New*

This analytical formula enables us to estimate the lowest possible value of ω_0/ω_u in practical situations. Experimentally, it is possible to make bilayer systems separated by a distance of d of the order of 10 nm. For rings of radius $R = 10$ mm and $a = 0.25$ mm, we find that $\omega_0 \approx 7 \times 10^{-5}\omega_u$, indicating that the longest resonance wavelength can be made of the order of 10^5 the radius of the ring if r_c can be ignored. This huge ratio is previously inaccessible. Experimentally both superconducting and ordinary SRR's have been studied. For nonsuperconducting rings, a detailed examination of Eq. (3.4a) under the three-mode approximation shows that our estimate is valid (i.e., ω_0 taking a non-zero real part) only when the resistivity is small enough so that $r_c < (4/3)\sqrt{L_0/C_1}\left(1 - C_1/C_1'\right)^{0.5}$. This implies that the total resistance of the ring, $r_c \cdot 2\pi R$, has to be approximately less than $Z_0 g\left(1 - C_1/C_1'\right)^{0.5}$, , where $Z_0 = (\mu_0/\varepsilon_0)^{1/2} = 377\Omega$ is the vacuum impedance, and $g = 2\sqrt{L_0/C_1}/3\pi R Z_0$ is a dimensionless constant, depending only weakly on a/R (for $a/R = 0.025$, we get $g = 6.26$). We next discuss the nature of the spectrum of the BC–SRR.

The spectrum of resonance frequency in the BC-SRR case is quite similar to that of a coplanar SRR as depicted in Figs. 3.2, 3.3, and each single-ring mode has again split into a pair of two modes with different symmetries. However, we found that the dipole moments exhibited by the resonance modes in a BC-SRR are quite *different* from those in the coplanar SRR (Fig. 3.3). In particular, in addition to the

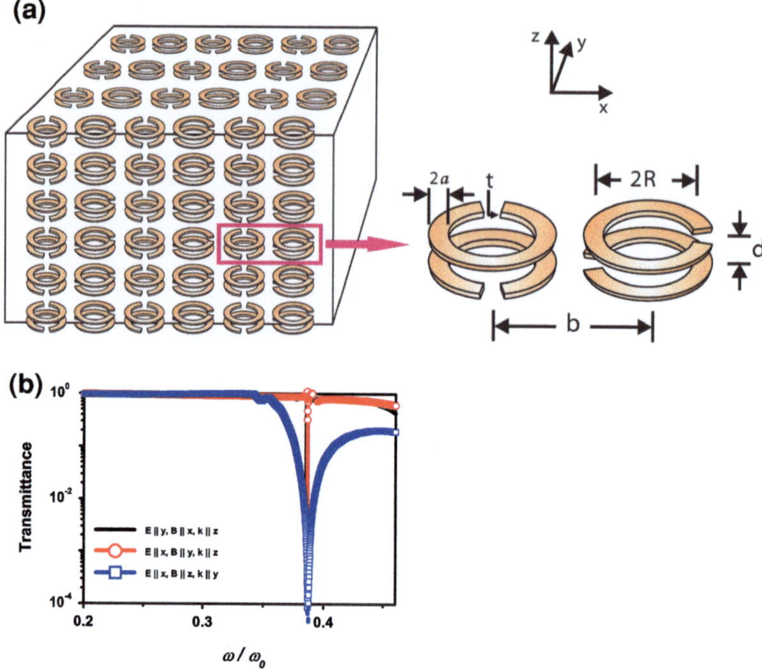

Fig. 3.10 **a** A schematic picture of the designed 3D magnetic material, with the inset showing the unit cell of structure. Here, $R = 4$ mm, $2a = 0.2$ mm, $t = 0.05$ mm, $d = 3$ mm, $b = 11$ mm and the gap width $\Delta = \pi/40$. (**b**) FDTD-calculated transmission spectra through the designed 3D magnetic material with different probing EM waves as specified in the figure. Reproduced from [8]

perpendicular magnetic polarization m_z, two more in-plane polarization (m_x, m_y) can also be induced for the BC-SRR under specific conditions. We summarized the characteristics of the EM polarizations for the lowest four modes in Table 3.1. Two important conclusions can be drawn. First, all the resonance modes in a BC–SRR are *completely free of bi-anisotropy* (i.e., either *purely* magnetic or *purely* electric). Second, some resonance mode (say, ω_1^L) could possess two magnetic moments simultaneously.

The second property of the BC-SRR enables us to design a *super* resonance unit for metamaterials. As we know, metamaterials that possess magnetic responses along all three dimensions drew much attention recently [4, 9]. As many resonant structures are inherently anisotropic, a standard method to design three-dimensional (3D) magnetic materials is to rotate the unite cell element and combine it with the original one to form an isotropic unit cell [4, 9]. Here, we provide an alternative approach. We demonstrate that a *layered* structure, composed of planar arrays of BC–SRR's, can exhibit magnetic responses along all three dimensions at the same frequency. Since the structure is basically a multilayer system, it is very easy to fabricate, particularly in higher frequency regime where the complex 3D structures are relatively difficult to fabricate. As shown in Fig. 3.10a, the unit cell of the designed metamaterial contains two BC–SRR's, with one rotated by 90 degrees with respect to the other one. To understand the resonance properties of the designed system, we employed FDTD simulations to calculate the transmissions of EM plane waves with magnetic fields polarized along different directions and with different propagation directions. The transmission spectra in different cases are compared in Fig. 3.10b. It is clearly shown that for **B** field along all three directions, a *common* resonance is excited at the frequency $\omega \sim 0.386\omega_u$, implying that the system exhibits strong responses to external magnetic fields along all three directions at this particular frequency.

3.5 Summary

In this chapter, we have extended the mode-expansion theory to more complex but realistic situations, in which the system can possess more metallic rings and in the metallic wire forming the ring does not necessarily exhibit a circular cross-section. The extended theory was applied to study the rich EM resonance properties of two practical systems, i.e., the coplanar double-ring SRR and the broadside-coupled double-ring SRR, yielding several intuitive analytical formulas which are helpful for practical designs. The validity of the extended theory was well justified by comparing with full wave numerical simulations and available experimental data.

References

1. J.B. Pendry, A.J. Holden, D.J. Robbins, W.J. Stewart, IEEE Trans. Microwave Theory Tech. **47**, 2075 (1999)
2. D.R. Smith, W.J. Padilla, D.C. Vier, S.C. Nemat-Nasser, S. Schultz, Phys. Rev. Lett. **84**, 4184 (2000)
3. K. Aydin, I. Bulu, K. Guven, M. Kafesaki, C.M. Soukoulis, E. Ozbay, N. J. Phys. **7**, 168 (2005)
4. J.D. Baena, L. Jelinek, R. Marqués, J.J. Mock, J. Gollub, D.R. Smith, Appl. Phys. Lett. **91**, 191105 (2007)
5. R. Marques, F. Medina, R. Raffi-El-Idrissi, Phys. Rev. B **65**, 144440 (2002)
6. R. Marques, F. Mesa, J. Martel, F. Medina, IEEE Trans. Antennas Prop. **51**, 2572 (2003)
7. L. Zhou, S.T. Chui, Appl. Phys. Lett. **90**, 041903 (2007)
8. X.Q. Huang, Y. Zhang, S.T. Chui, L. Zhou, Phys. Rev. B **77**, 235105 (2008)
9. Th. Koschny, L. Zhang, C.M. Soukoulis, Phys. Rev. B **71**, 121103 (2005)

Chapter 4
Helical Structures

4.1 Introduction

In this chapter we discuss metallic helixes. The helix is a typical example of a class of wire structure that is topologically the same as the split ring but does not possess circular symmetry. Different Fourier components of the current are coupled. The dipole antenna that consists of a single straight wire is another example of structure of lower symmetry. Because the dipole antenna is already so well known, we feel that the helix will be a more interesting example that illustrates the essential physics of this class of structures. There are two key points that we want to bring out in this chapter.

(1) Now the impedance matrix (contributed by inductance/capacitance terms) is no longer diagonal, the problem is still quite simple: Whereas the diagonal elements of this matrix are "log-divergent" and large, the off-diagonal elements are not and are usually smaller than the diagonal elements. *Thus the impedance matrix is often nearly diagonal.* Its inversion is simple and physically understandable.

(2) Even though the resonance modes, when expressed in terms of the arc length, are very similar to previous cases as is described in Chap. 2 and are well approximated by sinusoidal functions, the response is different because of the different shape of the wire structure. Thus the radiation direction can be axial or broadside, depending on the number of turns and the system parameters.

There are many practical examples for the helical structure. Macroscopic metallic helixes have been much exploited as antennae in microwave communication such as in GPS satellites. These usually involve helixes macroscopic in size and has many turns. The resonance frequency of interest is not the lowest one but is higher than the lowest one by a factor proportional to the number of turns. Experimental realization of helical structures on a nanoscale is becoming accessible and has been discussed by different groups [1–3]. The possibility of making helixes at micron scales enables the recent experimental study of these as efficient circular polarizers at the far infrared

S. T. Chui and L. Zhou, *Electromagnetic Behaviour of Metallic Wire Structures,*
DOI: 10.1007/978-1-4471-4159-4_4, © Springer-Verlag London 2013

[4] and as negative refracting material [5]. Helixes exhibit chiral symmetry. Micron size versions of these are prime candidates for artificial high frequency magnets. We have recently proposed [7] that micron size helixes may exhibit a giant (orders of magnitude larger than current materials) Faraday rotation at infrared frequencies. These small-scale structures exploit the low lying resonances. The microhelixes usually have only a few turns and the lowest few resonance modes are of interest.

This chapter focuses on the current $\mathbf{I}$ induced in a metallic helix by an external field $\mathbf{E}_{ext}$ which we denote symbolically as $\mathbf{I} = \mathbf{K}\mathbf{E}_{ext}$ where the kernel $\mathbf{K}$ is given in Eqs. (4.5) and (4.6) below. Quite often one is interested in the scattering properties of an object. This is captured by the t matrix, which relates the scattered field $\mathbf{E}_{sc}$ to the external field and is defined by the relationship: $\mathbf{E}_{sc} = \mathbf{t}\,\mathbf{E}_{ext}$. From the current in the helix, one can calculate the scattered field with the Green's function G, as is given in Eq. (0.3): $\mathbf{E}_{sc} = \mathbf{G}\mathbf{I}$. Thus the t matrix is given by $\mathbf{t} = \mathbf{G}\mathbf{K}$. This single helix t matrix makes it easier to calculate the T matrix for a collection of finite size helixes and hence the photonic band structure. This will be explained in more detail in Chaps. 8 and 9.

This chapter is organized in the following way. In the next section, we establish a general theoretical framework to compute the inductance/capacitance matrix for a metallic thin wire in arbitrary shape, by extending the theory presented in Chap. 2 (valid only for ring geometry) to general situations. We, then, describe the formalisms that we obtained for the helix structures and the behaviors of the circuit parameters for such a structure in Sects. 4.3 and 4.4.

4.2 Circuit Parameters for an Arbitrary-Shaped Metallic Thin Wire

As shown in Fig. 4.1, we consider a metallic thin wire with a circular cross-section and a length L, bent over to exhibit an arbitrary shape. Suppose there is a current flowing along the metallic wire described by the current density $\vec{j}(\vec{r'})$, then the inductive and capacitive fields can be calculated by putting $\vec{j}(\vec{r'})$ into Eqs. (1.4) and (1.5) to perform the integrations. Following the same argument as in Chap. 2, we assume that the current is localized at the center of the wire and calculate the inductive/capacitive fields at an arbitrary (convenient for calculations) position at the wire surface. To simplify the integrations, for each wire segment at the vicinity of a point $\vec{r}$, we introduce a local Cartesian coordinates system $\{\zeta, \eta, \xi\}$ with the coordinate origin at the wire center. Here, $\hat{e}_\xi$ is along the wire while $\hat{e}_\zeta, \hat{e}_\eta$ are the two orthogonal axes to define the local cross-section of the wire at a particular point. Therefore, the current density at the vicinity of the point $\vec{r'}$ must be written as

$$\vec{j}(\vec{r'}) = \hat{e}_{\xi'}I(\xi')\delta(\zeta')\delta(\eta')$$

and therefore, the inductive field can be evaluated as

Fig. 4.1 Schematic picture
of a general metallic thin wire

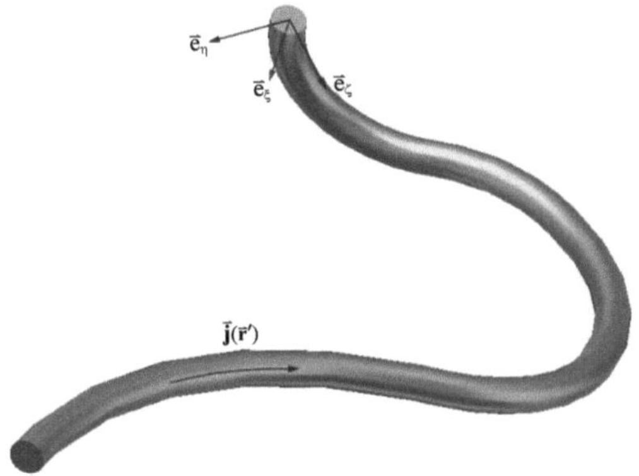

$$
\begin{cases}
\vec{E}_L(\xi) \cdot \hat{e}_\xi = -i\omega \displaystyle\int (\hat{e}_\xi \cdot \hat{e}_{\xi'})I(\xi')g(\xi,\xi')d\xi'/c^2 \\[3mm]
\vec{E}_C(\xi) \cdot \hat{e}_\xi = \dfrac{1}{i\omega} \displaystyle\int \dfrac{\partial I(\xi')}{\partial \xi'} \cdot \dfrac{\partial g(\xi,\xi')}{\partial \xi} d\xi'
\end{cases}
\tag{4.0}
$$

with the reduced Green's function defined as

$$
g(\xi,\xi') = G(r,r')|_{\varsigma=0,\,\eta=0;\,\varsigma'=\varsigma_0,\,\eta'=\eta_0}
$$

Here, $\{\varsigma_0, \eta_0, \xi\}$ denotes the position on the wire surface at which the electric fields are calculated. It is interesting to note that by setting $\xi = R\phi$ the formulas derived for the ring geometry are recovered.

We next represent these functions as Fourier series. The current is zero at the two ends: $I(0) = I(L) = 0$ and it is straightforward to represent it as a continuous periodic function. However, this condition does not necessarily hold for the electric field. Just like the split ring, there are, in general, two types of resonances: those of even or odd symmetry under a parity transformation about the midpoint of the helix. For the modes with odd symmetry we have no such obvious periodic condition. We can extend the argument ξ to $[0, 2L]$ by assuming $f(\xi) = 0, L < \xi < 2L$, and reassign the periodic condition $f(0) = f(2L)$. In such a doubled argument region, we are allowed to expand the physical quantity as:

$$
f(\xi) = \sum_m e^{imk_0\xi} f_m
$$

where the Fourier component should be calculated by:

$$
f_m = \frac{1}{2L}\int_0^{2L} f(\xi)e^{-imk_0\xi}d\xi = \frac{1}{2L}\int_0^{L} f(\xi)e^{-imk_0\xi}d\xi
$$

with $k_0 = \pi/L$. This approach is adopted in our general formulation and will be explained and illustrated in detail in the next two chapters. In this chapter, we focus on the modes with even symmetry. In that case $E(0) = E(L)$ where no such extension is necessary. These modes are the usual ones that are coupled to slowly varying external fields. In terms of the Fourier transform, we get from Eq. (4.0) that

$$\begin{cases} E_L^m = -i\omega \sum_m L_{mm'} I_{m'} \\ E_C^m = -\dfrac{1}{i\omega} \sum_{m'} (C^{-1})_{m,m'} I_{m'}. \end{cases}$$

where E_L^m, E_C^m are Fourier components of Eq. (4.0) and the circuit parameters are

$$\begin{cases} L_{mm'} = \frac{\mu_0}{2L} \iint (\hat{e}_\xi \cdot \hat{e}_{\xi'}) e^{ik_0(m'\xi' - m\xi)} g(\xi, \xi') d\xi' d\xi \\ (C^{-1})_{mm'} = -\frac{1}{2L\varepsilon_0} \iint (im'k_0) e^{ik_0(m'\xi' - m\xi)} \left(\frac{\partial}{\partial \xi} g(\xi, \xi') \right) d\xi' d\xi \end{cases}$$

Therefore, we can numerically calculate all the circuit parameters based on the above equations for a metallic wire of arbitrary shape.

4.3 Formalism for a Helix

We first describe the trajectory of a helix in cylindrical coordinates. The position of a point on the helix is given by:

$$x = (R\cos\phi, \ R\sin\phi, \ p\phi) \tag{4.1}$$

where the radius R and the pitch p are constants; the angle ϕ is considered as a parameter with a range $0 < \phi < 2\pi/\alpha$. Thus the "number of turns" $N = 1/\alpha$ determines the extent of the helix. The tangential vector is given by

$$\mathbf{v} = d\mathbf{x}/d\phi = (-R\sin\phi, \ R\cos\phi, \ p). \tag{4.2}$$

In terms of the circular basis set $\vec{e}_\pm = (\vec{e}_x \pm i\vec{e}_y)/\sqrt{2}$, $\mathbf{v}$ can also be written as $\mathbf{v} = -ie_+ \exp(-i\varphi)R/\sqrt{2} + ie_- \exp(i\varphi)R/\sqrt{2} + p\mathbf{e}_z$; $\mathbf{e}_x, \mathbf{e}_y, \mathbf{e}_z$ are unit vectors along the x, y, and z directions. The magnitude of this tangential vector is $v = (p^2 + R^2)^{1/2}$. For a helix with $N = 1/\alpha$ turns, the total arc length is $L = 2\pi N v$.

The current is along the tangential direction. We expand its magnitude as a Fourier series in terms of the arc length along the helix:

$$\mathbf{I}(\phi) = \mathbf{v} I(\phi)/|v| \tag{4.3}$$

where $I(\phi) = \sum_m I_m \exp(im\alpha\phi)$. We also expand the tangential component of the electric field in a Fourier series so that its mth component is given by: $E_m = \int d\phi \exp(-im\phi/\alpha) \mathbf{E} \cdot \mathbf{v} \alpha/(2\pi|v|)$.

As usual, this electric field contains contributions from the induced emfs from Faraday's law and from Coulomb's law as $\sum_m X_{m,m'} I_{m'}$; the impedance matrix is given by $X_{m,m'} = [iL_{m,m'}\omega - i/(\omega C_{m,m'})]$ where L, C are the effective inductances and the capacitances of the system:

$$L_{m,m'} = \int d\phi \exp(im\alpha\phi)R \int d\phi' \mathbf{v} \cdot \mathbf{v}' \exp(-im'\alpha\phi')g(\phi, \phi')/(2\pi\alpha c^2 v^2)$$

$$1/C_{m,m'} = mm' \int d\phi \exp(im\alpha\phi)R \int d\phi' \exp(-im'\alpha\phi')g(\phi, \phi')R. \qquad (4.4)$$

Here $\mathbf{v} \cdot \mathbf{v}' = \mathbf{R}^2 \cos(\varphi - \phi') + p^2$.

These circuit elements are no longer diagonal in m. For the capacitive term, $1/C_{m,m'} \propto mm'$, one factor of m is from relating the charge to the current, the other factor of m' is from relating the electric field $E_{m'}$ to the electrostatic potential $U_{m'}$. Similar to Eq. (2.18), in the presence of an external electric field, the current is determined by the circuit equation which, in terms of σ, the localized electric field at the ends, is given by

$$\sum_{m'} X_{m,m'} I_{m'} = E_{\text{ext}}(m) + \sigma.$$

Inverting this, we obtain

$$\mathbf{I} = \mathbf{X}^{-1}(E_{\text{ext}} + \sigma) \qquad (4.5)$$

From the constraint $I(s = 0) = \sum_m I_m = 0$, we obtain

$$\sigma = -\sum_{m,m'} X_{m,m'}^{-1} E_{\text{ext}}(m') / \sum_{m,m'} X_{m,m'}^{-1}. \qquad (4.6)$$

We first determine the resonance of the system, which is obtained when $E_{\text{ext}} = 0$. From Eq. (4.6) we arrive at the eigenvalue equation

$$\sum_{m,m'} X_{m,m'}^{-1} = 0. \qquad (4.7)$$

The corresponding eigenfunction I^α is given by

$$I^\alpha{}_m = J \sum_{m'} X_{mm'}^{-1} \qquad (4.8)$$

for some normalization constant J. Note that Eq. (4.7) involves a double sum over both m and m' whereas Eq. (4.8) involves only a single sum over m'. Near a resonance, the dominant contribution to the current comes from the second term on the right hand side of Eq. (4.5), which can be written approximately in a suggestive way as: $-|I^\alpha> <I^\alpha|E> / \sum_{m,m'} X_{m,m'}^{-1}$. Here the angular brackets represent a sum over the indicies of the vectors (inner product).

Now, $X^{-1} = X^c / \det(X)$ where X^c is the matrix of cofactors. Hence our condition for resonance becomes

$$\sum_{m,m'} X^c_{m,m'} = 0.$$

Because of the factor m, m' multiplying $1/C$ in X, the impedance matrix becomes big quickly as more Fourier modes are involved. The lowest resonance involves only a few Fourier modes, a good approximation for the lowest mode (Eq. 3.14) is given by

$$\mathbf{I} = I_0 \mathbf{v}[1 - \cos(\phi\alpha)]/|v|. \tag{4.9}$$

In this approximation,

$$2\hat{X} = \begin{bmatrix} \omega^2 L_1 - 1/C_1 & 2\omega^2 L_{1,0} \\ 2\omega^2 L_{1,0} & 2\omega^2 L_0 \end{bmatrix}$$

From Eq. (4.7) we obtain the resonance frequency

$$\omega_r^2 = [1/C_1]/[2L_0 - 4L_{1,0} + L_1].$$

where $L_1 = L_{1,1} + L_{1,-1}$, $\quad L_0 = L_{0,0}$, $\quad C_1 = C_{1,1} - C_{1,-1}$,

In the thin-wire limit when the off-diagonal impedance matrix elements are small (see below), the resonance modes are approximately the same as is discussed in the previous chapter. The resonance wavelengths are then

$$\lambda_n = 2L/n \tag{4.9a}$$

for integer n. The corresponding current distribution is given by

$$\vec{I}_n(\varphi) = \vec{v}\sin[(n + 1/2)\alpha\varphi]I/|v|. \tag{4.9b}$$

We next discuss the impedance parameters of the helix.

4.4 Circuit Parameters

In this section we discuss the circuit parameters in Eq. (4.4). Our aim is to explain why the matrix of parameters is mainly diagonal, the off-diagonal elements are usually smaller. For the even magnetic resonances with $I_m = I_{-m'}$ that we focus on here, one ends up with the symmetric combinations

$(L_{m,m'} + L_{m,-m'})/2 = l_{m,m'}/(2\pi\alpha c^2)$, $(C^{-1}_{m,m'} + C^{-1}_{m,-m'})/2 = \alpha c^{-1}_{m,m'}/(\pi R v)$ where the dimensionless parameters

$$c_{m,m'}^{-1} = mm' \int d\phi \, \sin(m\alpha\phi) \int d\phi' \, \sin(m'\alpha\phi') \exp(ik|r - r'|)R/|r - r'|,$$

$$l_{m,m'} = \int d\phi \, \cos(m\alpha\phi)R \int d\phi' \mathbf{v} \cdot \mathbf{v}' \cos(m'\alpha\phi')2\exp(ik|r - r'|)/|r - r'|.$$

$$(4.10)$$

For $m = m'$, these integrals are *log divergent* because the denominator of the integrand approaches zero when r approaches r'.

The off-diagonal impedance with $m \neq m'$ is not log divergent, however. This comes about because the off-diagonal components are controlled by the overlap integrals of orthogonal functions. For example, for $l_{m,m'}$ in Eq. (4.10), the log divergent contribution comes from contributions when r is close to r' (φ is close to φ'). This contribution is controlled by the integral $\int d\phi \, \cos(m\alpha\phi) \cos(m'\alpha\phi)$. This overlap integral is zero because the basis functions are orthogonal to each other. This result is true for all wire structures topologically the same as a single line, is true not just in the quasi-static limit. It is the crucial reason why, when expressed in terms of the arc length, the resonance modes of this class of wire structures are similar. This result is empirically known and has been exploited in the so-called meander structure antenna. The reason is never explicitly pointed out, however. The present approach also provides for the evaluation for the nondivergent contributions which distinguishes among the different geometries. This is one of the crucial simplifying features for wire structures. We provide next an example of the numerical values of these circuit parameters.

To illustrate the crucial physics, we ignore the variation of the current across the wire and consider the quasi-static limit. For finite size wires, there is a lower limit cutoff so that approximately, $|\mathbf{r} - \mathbf{r}'| > \mathbf{a}$ when the finite size of the wire of radius a that make up the helix is included.

We have computed the integrals in Eq. (4.10) numerically for different cutoff parameters $1/N$ with the constraint that $|\phi - \phi'|\alpha v > 1/N$. The results for the dimensionless components in the quasi-static limit are shown in Fig. 4.1 (symbols) for different values of the cutoff $1/N$. The numerical results for the diagonal components of the circuit parameters are fitted with a log term. We get

$$l_{1,1} = -13.53 + 12.28ln(N),$$
$$1/c_{1,1} = -27.67 + 12.32ln(N),$$
$$l_{2,2} = -36.95 + 12.266ln(N),$$
$$1/c_{2,2} = -38.57 + 12.32ln(N).$$
$$l_{0,0} = -53.69 + 24.6ln(N).$$

The results of the fit are shown as lines in Fig. 4.2. The off-diagonal components with $m \neq m'$ are down by two orders of magnitude and are not log divergent. For example, the off-diagonal inductance $-l_{1,0}$ is also shown in the figure (dotted line). In this figure, $-l_{1,0}$ is multiplied by a factor of 75 so that it can fit in the same graph .

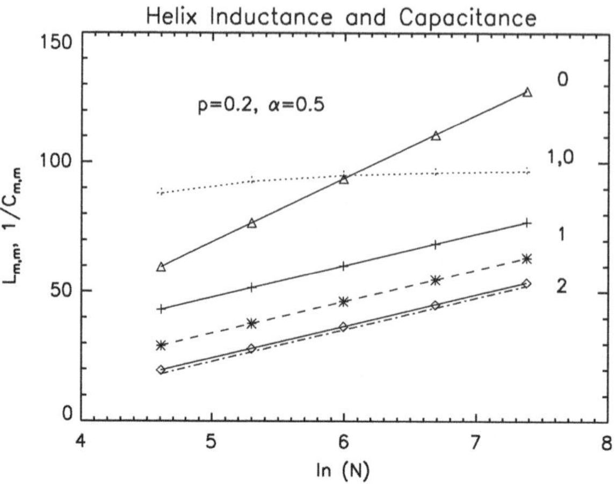

Fig. 4.2 Inductance (*solid lines* and corresponding points) and capacitance (*dashed* and *dotted-dashed lines* and corresponding points) of different angular momentum values as a function of the cutoffs. The points are the numerical results. The lines represent the logarithmic fits: The nearly horizontal *dotted line* is for $-75\,l_{1,0}$. {Reprinted with permission from Ref. [7]. Copyright (2008), American Institute of Physics.}

We have investigated the size dependence of these quantities and found that the coefficient of the log term scale linearly with $1/\alpha$. The impedances l and c are symmetric only in the quasi-static limit. We next discuss the response of this system.

4.5 Response

As is discussed at the beginning of this chapter, from the current in the helix, one can calculate the scattered field with the Green's function G, as is given in Eq. (2.3): $\mathbf{E_{sc}} = \mathbf{GI}$. This involves integrals of the current over the helix. There are basically two regime of interest, depending on the relative magnitude of the wavelength of the EM wave and the height of the helix. We discuss here the long wavelength limit. In that case, the scattered field can be obtained from the current induced by an external field by a multipole expansion. The discussion in this section is also applicable to the split rings discussed in previous chapters. Other cases will be discussed in the specific sections below.

Just like the split rings, the metallic helical structures are magnetoelectric in that an electric (magnetic) field can cause a magnetic (electric) polarization.

$$\mathbf{M} = \chi_{\mathbf{m}}\mathbf{B} + \chi\mathbf{E}, \quad \mathbf{P} = \beta_{\mathbf{e}}\mathbf{E} + \beta\mathbf{B}. \tag{4.11}$$

The coefficients χ and β are *anisotropic*. They are usually imaginary. This kind of behavior has been discussed numerically in Chap. 3. It is also well known from previous studies of optically active insulating helical structures, where the evaluation of the coefficients is carried out with perturbation theory. Here we are in the conducting "strong coupling" limit. As is emphasized at the beginning of this chapter, we wish to illustrate that the responses are different from the dipole antenna, even though the nature of the resonances are similar. We find that in the long helix limit, the coefficients are nearly diagonal whereas in the short helix limit they are not. We evaluate these coefficients with an analytic approach next.

The current induced by an external field is given in Eqs (4.5) and (4.6). From the induced current, one can calculate the induced multipole moments. To illustrate, consider the thin-wire limit with the current distribution given approximately by Eq. (4.9b). The magnetic dipole moment can be evaluated with the formula

$$\mathbf{m} = \int \mathbf{r} \times \mathbf{I}\, d\mathbf{r}/2c.$$

We find that

$$\mathbf{m} = \mathbf{f}R^3 I_0/(2|v|c). \tag{4.12}$$

$$f_x/p = \sin(2\pi/\alpha)\alpha'[1 - 2/(\alpha'^2 - 1)]/(\alpha'^2 - 1) - 2\pi(n + 1/2)\cos(2\pi/\alpha)/(\alpha'^2 - 1),$$
$$f_y/p = [1 + \cos(2\pi/\alpha)]\alpha'[1 - 2/(\alpha'^2 - 1)]/(\alpha'^2 - 1) - 2\pi(n + 1/2)\sin(2\pi/\alpha)/(\alpha'^2 - 1),$$
$$f_z = 2/\alpha'.$$

where $\alpha' = (n + 1/2)\alpha$, n is the mode index. (The usual factor of π is absorbed in the factor I_0, the extra factor of R is cancelled by the factor of $|v|$ in the denominator). The x, y components of the magnetic moment depend on the pitch p. This is similar to what is seen in the fishnet structure and the vertical rings discussed in Chap. 3.

The magnetic moment sometimes depends on the choice of the origin. If the origin is displaced by an amount r_0, the magnetic moment is changed by an amount $\Delta\mathbf{m} = \mathbf{r}_0 \times \int \mathbf{I} dr/2c$. If $<\mathbf{I}> = \int \mathbf{I} dr \neq 0$, this change is finite. For our case, we have $<\mathbf{I}> = -RI_0/|v|\alpha'\mathbf{h}$ where

$$h_x = \sin(2\pi/\alpha)/(\alpha'^2 - 1), \ h_y = -[1 + \cos(2\pi/\alpha)]/(\alpha'^2 - 1) \ h_z = -2p/\alpha'^2/R. \tag{4.13}$$

The charge density is related to the current density with the charge–current conservation equation: $\partial Q/\partial t + \nabla \cdot I = 0$. We obtain

$$Q = i\alpha' I_0 \cos(\phi\alpha')/(|v|\omega).$$

The corresponding electric dipole moment can be evaluated with the formula

$$\mathbf{p} = \int dr Q(r)\mathbf{r}.$$

We obtain

$$\mathbf{p} = \mathbf{h} i \alpha' I_0 R / (|v| \omega) \tag{4.13a}$$

where $\mathbf{h}$ has been given in Eq. (4.13). *The orientation of the dipole moment depends on the length of the helix.* For example, for a helix with one turn and for the lowest mode, $\alpha = 2\alpha' = 1$, $h_x = 0$, $h_y = 8/3$, $h_z = -8p/R$. The dipole moment is tilted in the zx plane. There is interest in designing GPS helical antenna so that the radiation can come either from the side (broadside) or the end (endfire). In general, electric dipole radiation has a maximum in a direction perpendicular to the orientation of the dipole. In this case, the corresponding radiation has an endfire component (from $p_{x,y}$) and a broadside component (from p_z). Similarly, for a helix with one and a half turns and for the lowest mode so that $n = 0$, $\alpha = 2\alpha' = 2/3$, we get $h_x = h_y = 0$, $h_z = -18p/R$. The electric dipole moment is along z. The corresponding electric dipole radiation is broadside. In general, the x, y components of the dipole moment are zero when $\alpha = 2/(2j + 1)$ for any integer j. Thus in this thin-wire approximation when the number of turns $N = j+1/2$ is a sum of an integer and an extra half turn, the radiation is broadside.

The denominator of $h_{x,y}$ becomes zero when $\alpha(n + 1/2) = \alpha' = 1$. This, however, is not the exact optimal condition for increasing the endfire radiation as then the numerator of $h_{x,y}$ also becomes zero. The extremum of $h_x^2 + h_y^2$ can be determined from the condition $4\alpha'(n + 1/2)(1 + \cos 2\pi/\alpha) + 2\pi(\alpha'^2 - 1)\sin(2\pi/\alpha)/\alpha^2 = 0$. Empirically, it seems that $\alpha \approx (n + 1/4)$ is a reasonable compromise.

Even though Eq. (4.9b) looks very similar to current distributions in other "simply connected" wires, the induced multipole moments depend on the shape of the wire. The different orientations of the dipole moment lead to radiation that can be axial or broadside, depending on the number of turns of the helix.

To calculate the susceptibilities, we need to know the value of I_0 as a function of the external fields. The response to external electric fields has been discussed for split rings. This can be applied here. From Eq. (1.27a) and for frequencies close to the lowest mode we get approximately

$$I_0 \approx {<}0|E{>} \, / \, {<}0|[X(\omega) - X(\omega_f)]|0{>} . \tag{4.13b}$$

We write ${<}0|[X(\omega) - X(\omega_f)]|0{>} = \omega/\xi_0$ where $1/\xi_0 = [L_{0,0} + (L_{1,1} + L_{1,-1})/2][1 - (\omega_r/\omega)^2]$. Hence, $\xi_0 = \omega I_0/{<}0|E{>}$.

With this we can now calculate the magnetic and the electric susceptibility of the system.

We first address the response due to an external magnetic field along the z axis. In principle, this can be obtained using Eq. (0.7), the magnetic field integral equation. For the present case, this can be derived by a simple argument. For B along z, there is an induced emf around the helix equal to

$$E_B = -i\omega R B \vec{e}_\phi/(2c) = -i\pi R B \vec{e}_\phi/\lambda, \tag{4.13c}$$

which couples to the $m = 0$ component of the current of the lowest resonance. The effective emf is proportional to R/λ, which can be small. However, for sub-wavelength rings, external electric fields cannot contribute to an $m = 0$ component of the emf. Thus the corresponding "noise" is also smaller. From Eq. (4.13b) $I_0 \approx E_B \xi_0 / i\omega$. The susceptibilities χ_m and β can be obtained by substituting the expression for I_0 into Eqs. (4.12) and (4.13) for the magnetic and electric dipole moments and dividing the expression by B and the volume V per ring. We obtain

$$\chi_{mzz} = -f\xi_0\pi/(2c^2\alpha), \quad \beta_{jz} = ih_j\alpha^2\chi_{mzz}c/(\omega v) \tag{4.14}$$

where $f = R^3/V$, V is the average volume occupied by a helix.

We next address the response due to an external electric field. For coupling to an external electric field E, the components of the field and the current are coupled as a dot product, as is indicated by the form of the energy $\int \mathbf{E} \cdot \mathbf{j}$. In component form, this is $\sum_{i=x,y,z} E_i \int j_i$. Substituting the resonance form for the current and carrying out the integral over the helix for each of the current components, we obtain $<0|E> = -0.5i \sum_{j=x,y,z} E_j\alpha'h_j$, the current is thus given by

$$I_0 \approx -0.5i \sum_{j=x,y,z} E_j\alpha'h_j\xi_0/\omega \tag{4.15}$$

Just like the electric dipole moment, this coupling depends on the length of the helix.

The susceptibilities β_e and α can be obtained by substituting the expression for I_0 into Eqs. (4.8) and (4.9) for the magnetic and electric dipole moments and dividing the expression by E. We obtain

$$\beta_{ejk} = -\alpha'\beta_{e0}h_jh_k, \quad \chi_{zk} = -\beta_{kz} \tag{4.16}$$

where $\beta_{e0} = -\pi^2\alpha\xi_0Rc^2/(V\omega^2)$. h is given Eq. (4.13a). The x, y components of h_j diverges when $\alpha' = \pm1$. This resonance corresponds to a mode so the circumference of one turn is approximately equal to the wavelength, similar to the result for a single split ring. This condition is consistent with recent results on long helixes[5].

Both the split ring and the helix are magnetoeletric. Yet, the helix is chiral and the split ring is not. The difference lies in the finite diagonal elements of the magnetoeletric coefficients for chiral systems. For nonchiral systems, the magnetoelectric coefficients are off-diagonal.

4.6 Helical Antenna

One can think of the interaction of the electromagnetic wave with the helix to consist of two steps: (1) the EM wave excites a current in the helix, (2) which then emits radiation. A helical antenna is a good example that illustrates the second step, especially when the wavelength is not that much longer than the helix size. There is much empirical knowledge about helical antennas. The effect of the susceptibilities on possible antenna applications was mentioned above. We shall try here to provide here a brief description from our point of view.

We assume that an external driving force generates a current in the helix. Depending on the frequency of this force, the current distribution is close to some resonance form, as is described above (Eq. 4.9b and 1.26a). We study the radiation from the current induced along the helix. The radiation fields can be obtained from the vector potential at position r far away from the helix given by [See also Eq. (0.2)]

$$A(r) = \mu_0 \exp(ikr)K/(4\pi r) \tag{4.17}$$

where

$$K = \int J(r') \exp(-i\vec{k} \cdot \vec{r'})d^3r'; \tag{4.18}$$

the vector $\mathbf{k} = \mathbf{u}k$, where u is a unit vector pointing towards the observation point, J is the current density.

Because the size of the helix is comparable to the wavelength, it is not a very accurate approximation to evaluate the integral (4.18) with the multipole expansion for this type of application. From (Eq. 4.3), the current distribution can be written as $I = v \sum_n c_n I_n(\varphi)$ with coefficients c_n in terms of the current distribution for the nth resonance $I_n(\varphi)$ as is approximately given in Eq. (4.9b):

$\vec{I}_n(\varphi) = \vec{v}I_n(\phi), I_n(\phi) = \sin[(n + 1/2)\alpha\varphi]I/|v|$. For the helix, the nth resonance provides for a contribution to Eq. (4.18) given by

$K_n = \int (-R\sin\varphi, R\cos\varphi, p)I_n(\phi) \exp\{-i[k_\perp R\cos(\varphi - \varphi_u) + pk_z\varphi]\}d\varphi/v$

This integral can be evaluated using the identity $\exp(ix\cos\varphi) = \sum_q i^q J_q(x) \exp(iq\varphi)$.

We get:

$$K_n = \sum_q J_q(k_\perp R)i^{q+1} \int [-e_+ \exp(-i\varphi)R + e_- \exp(i\varphi)R + pe_z]I_n(\varphi)$$

$$\exp[-ipk_z\varphi + iq(\phi - \varphi_u)]d\varphi/v\sqrt{2} \tag{4.19}$$

The first square bracket on the right contains contributions from currents along different directions. The current along the z direction give rise to broadside radiation, while the current in the xy plane give rise to the "end-fire" radiation. We first look at the end-fire mode:

4.6.1 End-Fire Mode

For the axial mode end-fire antenna, the radiation moves in a direction along the cylinder axis. For antenna design, the empirical rule for the radius R and pitch p of a helix for a EM wave with wavelength λ is: $R \approx \lambda/(2\pi)$. The resonance wavelengths in the thin-wire approximation are $\lambda_n = \lambda_0/n$, where the largest resonance wavelength (angular frequency) is $\lambda_0 = 2\pi R/\alpha$. So, the empirical resonance of interest corresponds to the condition $\alpha' = (n + 1/2)\alpha \approx 1$ near the divergence of the h's discussed in Sect 4.5. In general, the helix do not have exactly an integer number of turns. We write $1/\alpha = n_0/2 + \delta$ where δ is a number with magnitude between 0 and 0.5.

For the end-fire mode, the wave vector is along z. We are interested in the term with the $\mathbf{K}_n$ in Eq. (4.19) along the xy plane. In the thin-wire limit, for the nth resonance mode with the current given in Eq. (4.9b). For the end-fire mode, $k_z = k$. Under this condition, the z component of $\mathbf{K}_{n0}$ is small. The dominant contribution to the xy component of the integral $\mathbf{K}_{n0}$ comes from the $q = 0$ term in Eq. (4.19). The total exponent in the integral is $i\varphi[kp \pm n/(n_0 + 2\delta)] - 1$ for the two circularly polarized modes. So the optimum pitch is given by the condition $[kp \pm n/(n_0 + 2\delta)] - 1 = 0$. Assume that δ is small. This condition becomes $kp = \pm n/n_0(1 - 2\delta/n_0) - 1$. This condition can be satisfied by $n = n_0$ and $kp = \pm 2\delta/n_0$. From this we obtain $2\pi p = 2\delta\lambda/n_0$. This differs slightly from the empirically rule that $2\pi p = \lambda/4$. The effect of damping has not been included in the above consideration.

4.6.2 Broadside Mode

For this mode, $k_z = 0$. Again, we focus on the term $q = 0$ in Eq. (4.19). The z component of $\mathbf{K}_{n0}$ is of the order of the integral of the current I_n:

$K_{nz} \sim pJ_0(kR) \int I_n(\varphi)d\varphi/v\sqrt{2}$. This is finite for odd-number modes $n = 2n'+1$ for some integer n'. The smaller n is, the larger the corresponding contribution.

4.7 Helical Circular Polarizer

In this section we consider an example of how a current is excited. Micron scale helixes have recently been made by several groups. A particularly impressive application is as absorbers for circular polarization. In these studies, the frequency corresponds to the lowest few resonance modes. The wave vector is parallel to the helix axis. The coupling of the external EM field to the nth resonance mode is the central quantity of interest.

This coupling is given by $<n|E> = \int_0^{2\pi/\alpha} I_n(\varphi)v \cdot E_{ext}d\varphi$. Again, $I_n(\varphi)$ is approximately given in Eq. (4.9b). The external field is of a form $E_{ext} =$

$(e_x \pm ie_y) \exp(ikz)$. The two signs are for left and right circularly polarized light. The details of the coupling matrix element are described the Appendix. Here we summarize our results.

Substituting in the value of the tangential vector, we obtain: $<n|E> = \pm iR \int_0^{2\pi N} I_n(\varphi) \exp[i(kp \pm 1)\phi]d\varphi$ where $I_n(\varphi) \approx \sin(n\varphi/2N)$. This is essentially the integral K_n in Eq. (4.19). The largest contribution to this integral $<n|E>$ comes from the resonance characterized by an index $n_1 \sim 2N(kp \pm 1)$ for a helix with N turns. For example, $N= 2$, $n_1 = 4(kp \pm 1)$. For $kp = 1/4$, we get $n_1 = -3, 5$. The experiment was carried out off-resonance so there is a broad range frequency response.

4.8 Faraday Rotation

In this section, we explore a consequence of the magnetoelectric properties of the helixes. In magnetic materials, left and right circularly polarized light travel at different velocities. This leads to the Faraday effect, which has been exploited in the design of different fundamental devices such as isolators. At frequencies above the spin wave frequency, the Faraday rotation is rather small. Isolators are typically macroscopic in size at high frequencies, of the order of centimeters. Sometimes an external magnetic field is required. Helixes are chiral in nature. Chiral molecules are known to be optically active and exhibit a Faraday rotation. But the effect exhibited is small and has been treated theoretically by perturbation theory with respect to the dipole operator. The metallic structures here are in the "strong coupling" limit instead and cannot be treated in the same way.

Near the resonance of this structure, the Faraday rotation can be very big. Since the resonance frequency is controlled by the size of the helix, the operating frequency can be much higher than that of magnetic materials. This opens the door to construct basic devices such as isolators on a nanoscale and incorporate them in a chip.

Terbium gallium garnet (TGG) is supposed to have an extremely high Faraday rotation. Its Verdet constant is approximately 40 radian/Tesla/meter. So for a field of 10 Tesla, the rotation rate is approximately $4 \times 10^{-4} radian/micron$, three orders of magnitude smaller than that produced by a micron size helix. This may be useful for isolators and circulators.

In Chap. 9, we discuss the propagation of EM waves through a uniform magneto-electric materials. Here we consider a periodic array of helixes with their axis aligned along the z axis with the light (wave vector k and angular frequency $\omega = ck_0$) going along the x direction. We shall treat our array as a uniform system with some effective susceptibilities. To illustrate the essential physics, we make the simplest approximation so that the susceptibility is approximately the average value of that of the matrix and the helixes given in Sect. 4.4, weighted by the corresponding volume fractions. An improved estimate can be obtained with the

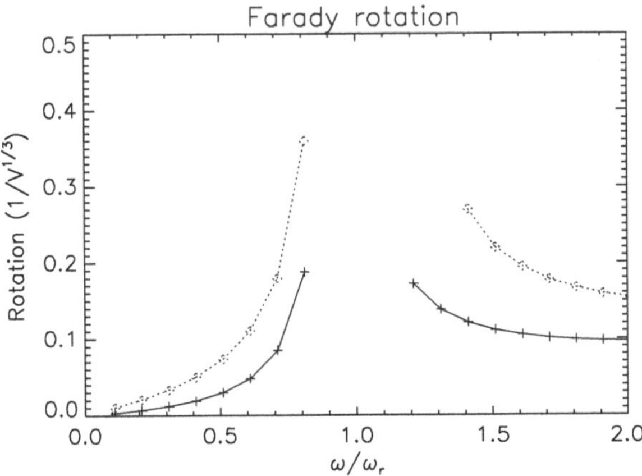

Fig. 4.3 Rotation rate in units of $V^{-1/3}$ as a function of the frequency normalized by the resonance frequency for $\alpha = 0.5$, (1) $p = 0.2$ (*solid line*); (2) $p = 0.4$ (*dotted line*). (Reprinted with permission from Ref. [7]. Copyright (2008), American Institute of Physics)

Claussius-Mosotti approximation or variations thereof. Because the helixes are chiral, they possess a finite diagonal magnetoelectric coefficient. We also focus on helixes with an extra half integer turn so that only α_{zz} is nonzero. From solving Maxwell's equation for this system, (see Chapter 8) we find that the speed of light, $u = \omega/k$, for the two circularly polarized modes is given by:

$$2\mu^{-1}(u_\pm/c)^2 = b \pm [b^2 - 4\mu^{-1}\varepsilon_{yy}\varepsilon_{zz}]^{1/2},$$

with the polarization given by:

$$E_z/E_y = -k_0 k \chi'^2 + k_0^2,$$

where $\varepsilon = 1 + 4\pi\beta_e$, $\mu^{-1} = 1 - 4\pi\chi_m$, $\chi' = 4\pi\chi\sqrt{1/\varepsilon}$, $b = (\mu^{-1}\varepsilon_{yy} + \varepsilon_{zz} - \chi'^2)$. The magneto-electric coefficient χ and hence χ' is imaginary. E_z, E_y are out of phase; the light becomes elliptically polarized in the transverse direction. The speed of light $u_\pm$ is significantly different for the two polarizations. This difference in velocity leads to a giant Faraday effect. (Fig. 4.3)

4.8.1 Numerical Estimates

Substituting in our estimate for the circuit parameters, we obtain an estimate of the resonance frequency in a "dimensionless" form: $\omega_r = \alpha\omega_0 w$ where $w^2 = (1/c_1)/(2l_0 - 4l_{1,0} + l_1)$, $\omega_0^2 = c^2/Rv$. We have estimated the rate of rotation with distance of the electric field for a linearly polarized plane wave from the difference of the phase velocity of the left and the right polarized modes given in Eq. (4.10). The result, in units of $V^{-1/3}$, is shown in Fig. (4.2).

Near the resonance, one of the modes is nonpropagating, hence the rotation rate is not shown. Thus if R and hence V is small, the rotation rate is very fast. For example, for a volume per ring V of micron^3, we get a rotation of $0.1 \text{radian}/\mu$.

Appendix

In this appendix, we describe the calculation of the coupling of an external field to the eigenmodes of the helix. The external field can be written in a general form as $E_{ext} = (a_- e_+ + a_+ e_- + a_z e_z) \exp(ik \cdot r)$ where $a_\pm = (a_x \pm i a_y)/\sqrt{2}$. For example, for $\boldsymbol{E}_{ext}$ in the xy plane and perpendicular to $\boldsymbol{k}$, $E_{ext} \propto (-\sin \varphi_k, \cos \varphi_k, 0)$. Then $a_\pm = -\sin \varphi_k \pm i \cos \varphi_k = \pm i \exp(\pm i \varphi_k)$

Recall that $I(\phi) = vI(\phi)/|v|$ with the tangent vector $v = \sum \mp i e_\pm \exp(\mp i\varphi)R/\sqrt{2} + pe_z$. The projection of the external field on the tangent vector can thus be written as:

$$E_{ext} \cdot v = \exp(ik \cdot r)\left[\sum \mp i a_\pm \exp(\mp i\varphi)R/\sqrt{2} + a_z p\right]$$

We obtain: $<n|E> = <n|E>_+ + <n|E>_- + <n|E>_z$ where

$$<n|E>_\pm = \mp a_\pm iR \int_0^{2\pi/\alpha} I_n(\varphi) \exp[i(k_z p\varphi + k_\perp R\cos(\varphi - \varphi_k) \mp \varphi)]d\varphi$$

$$<n|E>_z = a_z p \int_0^{2\pi/\alpha} I_n(\varphi) \exp[i(k_z p\varphi + k_\perp R\cos(\varphi - \varphi_k))]d\varphi$$

Using the identity $\exp(ix\cos\varphi) = \sum_q i^q J_q(x)\exp(iq\varphi)$, we get

$$<n|E>_\pm = \mp a_\pm iR \sum_m i^m \int_0^{2\pi/\alpha} I_n(\varphi)J_m(k_\perp R) \exp[i(k_z p\varphi + m(\varphi - \varphi_k) \mp \varphi)]d\varphi$$

In the thin-wire limit $I_n(\varphi) = \sin(n\varphi/2N)$. The dominant contributions are from the lowest m's:

$<n|E>_\pm = \mp a_\pm iR[\int_0^{2\pi N} I_n(\varphi)J_0(k_\perp R) \exp[i(k_z p\varphi \mp \varphi)]d\varphi$

$+i^{\pm 1} \int_0^{2\pi N} I_n(\varphi)J_{\pm 1}(k_\perp R) \exp[i(k_z p\varphi \pm (\varphi - \varphi_k) \mp \varphi)]d\varphi]$. This can be simplified as

$$<n|E>_\pm = \pm a_\pm R[J_1(k_\perp R)P_\mp \exp(\mp i\varphi_k) - J_0(k_\perp R)Q_\mp]$$

where $P_\mp = \int_0^{2\pi/\alpha} I_n(\varphi) \exp[i(k_z p\varphi \mp \varphi_k)]d\varphi$

$$Q_{\mp} = i\left[\int_0^{2\pi/\alpha} I_n(\varphi)\exp[i(k_z p\varphi \mp \varphi)]d\varphi\right]$$

For perpendicular incidence, $k_\perp = 0$, only the Q term contribute. This is the result quoted above. The $m = 1$ contribution depends on the angle of the wave vector. This coupling depends on the direction of the incoming wave vector. The coupling is optimal if, for example, $k_z p + n/(2N) = 1$.

For linear polarization along x, $a_+ = a_- = 1/\sqrt{2}$ we get the factor

$$iR[a_+ J_{-1}\exp(i\varphi_k)/i - a_- iJ_1\exp(-i\varphi_k)] = RJ_1[\exp(i\varphi_k) + \exp(-i\varphi_k)]/\sqrt{2}$$
$$= -RJ_1 i\sin(\varphi_k)\sqrt{2}$$

For polarization along y, $a_+ = -a_- = 1/i\sqrt{2}$ we get the factor

$$iR[a_+ J_{-1}\exp(i\varphi_k)/i - a_- iJ_1\exp(-i\varphi_k)] = -iRJ_1[\exp(i\varphi_k) + \exp(-i\varphi_k)]/\sqrt{2}$$
$$= -RJ_1 i\cos(\varphi_k)\sqrt{2}$$

References

1. K. Robbie, J.C. Sit, M.J. Brett, J. Vac. Sci. Tech. **B16**, 1115 (1998)
2. Y. K. Pang, J.C.W. Lee, H. F. Lee, W. Y. Tam, C. T. Chan, P. Sheng, Opt. Express **13**, 7615, (2005) and references therein.
3. R. Abdedda, G. Guida, A. Priou, B. Gallas, J. Rivory, Negative permittivity and permeability of gold square nanospirals. Appl. Phys. Lett. **94**, 081907 (2009)
4. J. K. Gansel et al., Gold helix photonic metamaterials as broadband circular polarizer. Science **325**, 1513 (2010)
5. J. B. Pendry, Science **306**, 1353 (2004)
6. C. Wu, H. Li, Z. Wei, X. Yu, C. T. Chan, Phys. Rev. Lett. **105**, 247401 (2010)
7. S. T. Chui, Giant wave rotation for small helical structures. J. Appl. Phys. **104**, 013904 (2008).

Chapter 5
General Multiply Connected Metallic Wire Networks: T and H

5.1 Introduction

In this chapter, we discuss the situation of general wire structures consisting of more than a single wire. There are many examples of this type of systems. A classic example is the Yagi, or fishbone antenna, which consists of parallel wires of different lengths and separations. Another example is a fractal structure of wires of different lengths which was studied as broadband absorbers [1, 2]. Recent foci are on wires of much smaller dimensions. A collection of straight wires can often mimic the physical properties of rings and helixes and exhibit strong magnetic responses. They are much easier to make on a nanoscale. A recent example is the fishnet structure (see the discussions in Chap. 2). Actually, the Yagi antenna also exhibits magnetic responses (S. T. Chui, unpublished data). A small scale Yagi antenna has recently studied [3].

For a collection of wires, there is a new topology where more than two wires can meet at a junction. The current is conserved at each junction. This current conservation condition can be incorporated with the introduction of local electric field not just at the free ends but at the junctions, as we demonstrate below. In this chapter, we shall explain the full equivalent circuit theory for finite frequency non-uniform currents in wire structures. This equivalent circuit theory amounts to a set of equations stipulating the relationship between the tangential electric field along the wire and the currents in the wires. In previous chapters we discuss the case of singly connected metallic wires. Here we study the extension to multiply connected wire networks. In the zero frequency limit with uniform currents, the behavior of currents in a network is governed by Kirchoff's laws. The consideration of multiply connected networks forces us to consider how to generalize the well-known Kirchoff's law. This chapter describes how this can be implemented [4]. A general theory for arbitrary wire structure is described. Simple examples are given for the T and the H shape structure. It turns out that there is a simple physical way to understand the nature of different resonances. This interpretation is explained.

S. T. Chui and L. Zhou, *Electromagnetic Behaviour of Metallic Wire Structures*,
DOI: 10.1007/978-1-4471-4159-4_5, © Springer-Verlag London 2013

C. Qu and L. Zhou (QZ) (unpublished data) have recently performed FDTD simulations to study the resonance properties of a series of T- and H-shaped structures. We found excellent agreement between their FDTD results and our approach. This is summarized in the last section.

5.2 Formulation

Although the boundary conditions on wire currents are essentially dictated by the continuity condition, i.e. currents diminish at the wire ends and sum of currents entering a joint equals the sum of currents leaving the joint, because the currents are no longer spatially uniform, the sum of the spatial derivatives of the currents at a junction is not zero. As a result localized charges and electric fields are introduced at the junctions. How to impose the boundary conditions on the wire ends as well as on wire joints is a nontrivial issue. While this problem can be avoided in the previously studied single-wire or non-crossing double-wire cases by appending a huge resistance at wire ends, that treatment cannot be extended to general wire networks with joints. Nevertheless, an important clue can be found by inspecting the results of previous chapters where the equation set can be easily reformulated into a Hermitian form with localized end electric fields instead of introducing large end electric wire resistances. To maintain this property for the generalized resonance equation set with boundary condition at wire ends and joints, localized boundary electric fields have to be taken into consideration. One can alternatively consider focusing on the localized charges at the ends instead of the fields but that formulation is not Hermitian. The generalized circuit equation set allows for calculating the response of the structure to an external field, as well as for a self-consistent determination of both current profiles and "boundary" electric fields for all the eigenmodes of a wire network.

Within the equivalent circuit theory, one seeks the relationship between the wire currents and the tangential components of electric field along the wires. The wire radius a_i comes into play since the tangential field is taken on the wire surface. To extract the inductances and capacitances of the wire network, we expand both the electric fields and the currents with a complete set of orthogonal functions. For the problem under consideration, the plane wave expansion is a good choice for each segment of the metallic wire. Any function $X(x)$ can be expanded as

$$X(x) = \sum_h X(h) \exp\left(+\frac{i\pi h x}{d}\right), \tag{5.1}$$

with the inverse transform

$$X(h) = \frac{1}{2d} \int_{x_1}^{x_2} X(x) \exp\left(-\frac{i\pi h x}{d}\right) dx. \tag{5.1a}$$

Here x_i is the two end of a wire segment, h is an integer, and $x_2 - x_1 = 2d$ where d is the wire length. Different from the single wire case, the currents at the two ends of a wire segment are no longer symmetrically related to each other. In the Fourier expansion, we have doubled the period to take this into consideration. The issue of padding is well-known in algorithms for fast Fourier transforms. There are different ways by which the function X can be extended to this new region. If one is careful, the result will be the same. One way is to set X to be symmetrically related to that in the original domain: $X(x - d) = X(2d - x)$ for $x > d$. We have chosen to set the current and the charge densities to be zero from $x_2 - d$ to x_2. Thus the range of integration in Eq. (5.1a) is then only of size d for those functions. If only the components with even h are nonzero, we retrieve the result of the previous sections of a single wire. With this choice, there are some mathematical issues which fortunately do not invalidate the final answer. We shall discuss this at the end of this chapter.

As usual the "inductive" and the "capacitive" electric fields can be recast into a matrix form in terms of the wire currents and impedances

$$E_L^i(h_i) = \sum_{j,h_j'} E_L^{ij}(h_i, h_j')I_j(h_j'),$$

$$E_C^i(h_i) = \sum_{j,h_j'} E_C^{ij}(h_i, h_j')I_j(h_j').$$

The Fourier transform of the circuit parameters $E_L^{ij}(h_i, h_j')$ and $E_C^{ij}(h_i, h_j')$ are defined in terms of the reduced inductances $L^{ij}(h_i, h_j')$

$$E_L^{ij}(h_i, h_j') = -\frac{i\omega}{2d_i c^2} L^{ij}(h_i, h_j')\hat{e}_i \cdot \hat{e}_j, \tag{5.2a}$$

$$E_C^{ij}(h_i, h_j') = -\frac{1}{2d_i i\omega} \left(\frac{\pi h_i}{d_i}\right)\left(\frac{\pi h_j'}{d_j}\right) L^{ij}(h_i, h_j'). \tag{5.2b}$$

Here $i = 1, 2, 3, \ldots$ is the index for the wires, $\hat{e}_i$ is a unit vector along ith wire. Equation (5.2a) suggests that a capacitance generally exists between any two wire segments, but the mutual inductance vanishes between any two wire segments which are perpendicular to each other. It is simple to check that $L^{ij}(h_i, h_j')$ is a Hermitian matrix in the quasi-static limit with no damping if the wire thicknesses are all taken to be the same.

The circuit parameter $L^{ij}(h_i, h_j')$ is equal to the integral $\int d\vec{r}\, d\vec{r}' G(r - r')$ $\exp[i\pi(hx - h'x')/d]$ where G is defined in Eq. (1.2). Just as we emphasized in the previous chapter, G becomes infinite as $r - r'$ approaches zero. For the diagonal term with $h = h'$, this integral is of the order of log (radius of wire/length of wire), logarithmically divergent in the thin wire limit. For h not equal to h', the circuit parameter $L^{ij}(h, h')$ is no longer log divergent because the basis functions $\exp(ihx)$ with different h are orthogonal to each other. The divergent contribution to $L^{ij}(h, h')$

is controlled by $\int_0^{2d} dx\,\exp[i\pi x(h - h')/d] = 0$. Thus the circuit parameters matrix is approximately diagonal. The circuit equations become simpler and its physical meaning is easier to understand. This is one of the advantages of our approach.

After substituting the inductance field (Eq. 5.2a) and the capacitance field (Eq. 5.2b) into the circuit equation (Eq. 1.1), we obtain the following equivalent circuit equation:

$$\frac{2cd_i^2}{i\pi\Omega_i}E(h_i) = \frac{2cd_iR_i}{i\pi\Omega_i}I_i(h_i) + \sum_{j,\,h_j'}\left[\hat{e}_i \cdot \hat{e}_j - \frac{h_ih_j'}{\Omega_i\Omega_j}\right]L^{ij}(h_i,\,h_j')I_j(h_j'). \tag{5.3}$$

This can also be written in a symbolic form as

$$\boldsymbol{E} = \boldsymbol{XI} \tag{5.3a}$$

where the impedance matrix is given by

$$\frac{2cd_i^2}{i\pi\Omega_i}X_{i,j} = \frac{2cd_iR_i}{i\pi\Omega_i}\delta_{i,j} + \left[\hat{e}_i \cdot \hat{e}_j - \frac{h_ih_j'}{\Omega_i\Omega_j}\right]L^{ij}(h_i,\,h_j'). \tag{5.3b}$$

$\Omega_i = \omega d_i/\pi c$ denotes the reduced frequency and $R_i = \rho d_i/\pi a_i^2$ is the resistance of the ith wire. In the absence of resistances and external fields, the eigenvalue equation for determining the resonances is Hermitian in the quasi-static limit, thus dissipationless modes are guaranteed.

A numerical issue of importance is the number of Fourier modes that we need to keep. The last term of the above equation increases quadratically as the Fourier mode indexes h, h' increase in magnitude. Usually for low lying excitations, only a few Fourier modes are necessary to produce accurate results. This is another advantage of our formulation.

Finally, a crucial issue is to take care of the boundary conditions at the ends and the junctions of the network. The boundary conditions are: (1) No current flow at the vertex of a wire (labelled by i) with an open end: Here we pick the origin not at the open end. In terms of the Fourier transforms, this can be written as

$$\sum_{h_i}(-1)^{h_i}I_i(h_i) = 0.$$

(2) Current conservation at a vertex that is not open:
All the boundary conditions lead to a set of equations of a form

$$\boldsymbol{H_0I} = 0 \tag{5.4}$$

for some matrix $\boldsymbol{H_0}$. Localized electric fields at the ends of the wires were discussed in previous chapters. We found that the simplest way to implement the current conservation constraints is to introduce localized electric fields at the junctions. These fields are essentially Lagrange multipliers to facilitate the introduction of the boundary conditions at the vertices of the wire network. The total electric field is then the sum of the external fields and the boundary fields:

$$E(h) = E_{\text{ext}}(h) + \sigma_{\text{b}}(h) - \sigma_{\text{e}}(h).$$

Here $\sigma_{\text{b}}(h)$, $\sigma_{\text{e}}(h)$ are the Fourier components of the electric fields at the *two* ends of a wire, which are now different. We have used a scalar symbol for this field to indicate that for each of the wires connecting to a vertex, the tangential components away from the vertex are all the same.

The circuit Eq. (5.3a) finally becomes:

$$\boldsymbol{E}_{\text{ext}} = \boldsymbol{XI} - \sigma_{\text{b}} + \sigma_{\text{e}} \tag{5.5}$$

where $\boldsymbol{X}$ is the impedance matrix. Because of the mutual capacitances and inductances between the wires, the currents on all the wires are coupled. From this we obtain

$$\boldsymbol{I} = \boldsymbol{X}^{-1}\boldsymbol{E}_{\text{ext}} - \boldsymbol{Y}^{-1}\sigma \tag{5.6}$$

where

$$Y_n^{-1}(h) = \sum_{n',h'} X_{n,n'}^{-1}(h, h'). \tag{5.7}$$

The subscript n corresponds to the wire index. This sum over n occurs for Y because σ is labeled by a vertex index and not a wire index. There is a sum over h' for Y in Eq. (5.7) because the electric field σ is localized and not a function of it. The boundary condition of current conservation at the vertices has been written in Eq. (5.4):

$$\boldsymbol{H}_0\boldsymbol{I} = 0. \tag{5.8}$$

where the matrix $\boldsymbol{H}_0$ is determined by the connectivity of the wire network. Examples of these are illustrated below. Substituting in the expression for the current, we obtain

$$\boldsymbol{H}_0\big(\boldsymbol{X}^{-1}\boldsymbol{E}_{\text{ext}} - \boldsymbol{Y}^{-1}\sigma\big) = \boldsymbol{0}. \tag{5.9}$$

Hence

$$\sigma = \sum_h (\boldsymbol{H}_0\boldsymbol{X}^{-1})_h \boldsymbol{E}_{\text{ext},h} \Big/ \sum_h (\boldsymbol{H}_0\boldsymbol{Y}^{-1})_h$$

We finally obtain

$$\boldsymbol{I} = \boldsymbol{X}^{-1}\boldsymbol{E}_{\text{ext},h} - \boldsymbol{Y}^{-1}\sum_h (\boldsymbol{H}_0\boldsymbol{X}^{-1})_h \boldsymbol{E}_{\text{ext},h} \Big/ \sum_h (\boldsymbol{H}_0\boldsymbol{Y}^{-1})_h \tag{5.10}$$

Resonance occurs when there is a nontrivial solution to the above in the limit of small external fields. Thus, the resonance condition becomes

$$\boldsymbol{H}_0\boldsymbol{Y}^{-1}\sigma = \boldsymbol{0}. \tag{5.10a}$$

Examples of these kinds of equations for different structures are shown below and in the next chapter.

The resonance frequency appears as a nonlinear parameter in the above equation. The condition (5.10b) implies that the determinant of $H_0 Y^{-1}$ is equal to zero. Numerically the value of the determinant is usually very large and changes rapidly. It is not easy to directly focus on the determinant. To determine these resonance frequencies, we generalize a technique previously used in the solution of the multiple scattering problem where a similar type of equation occurs. In general, for any matrix $M(\omega)$, to determine the frequencies ω so that $\det[M(\omega)] = 0$. We examine the number of eigenvalues with positive real parts of the matrix M for a distribution of frequencies ω. When this number is changed, that indicates that one of the eigenvalues may have crossed zero. The determinant of M, which is equal to the product of the eigenvalues, has also crossed zero. A resonance is indicated. There is one anomalous situation when an eigenvalue approaches positive infinity, then switches to negative infinity without crossing zero. Care should be taken to eliminate this case. This approach will be described in more detail below.

The present formalism not only self-consistently determines both the current patterns and the boundary fields in a wire network, but also can be easily extended to more complex network and facilitate the analysis of eigenmodes in a general metallic wire network.

Once the eigenvectors are available, it is straightforward to calculate the wire currents and the charge distributions. Taking into consideration that the currents and charges are both real numbers and only standing wave solutions are possible in a finite structure, the current and charge distributions are generally expressed in terms of current components as

$$I_i(x) = \mathrm{Re} \sum_{h_i} I_i(h_i) \exp\left(\frac{i\pi h_i x}{d_i}\right), \tag{5.11}$$

$$Q_i(x) = -\frac{1}{\Omega_1 c}\frac{d_1}{d_i} \mathrm{Im} \sum_{h_i} h_i I_i(h_i) \exp\left(\frac{i\pi h_i x}{d_i}\right). \tag{5.12}$$

We discuss below some simple applications of the above formulation.

5.3 The "T" Structure

With the formalism outlined above, to illustrate our method, we studied two examples of wire networks in this chapter: the T-shape and H-shape wire structures [4]. The T structure is probably the simplest wire network with a joint and demonstrates the essential features of the resonance modes. For the H-shape wire network, a rich variety of current configurations are available which exhibit comparable electric or magnetic responses. Detailed analysis shows that the type

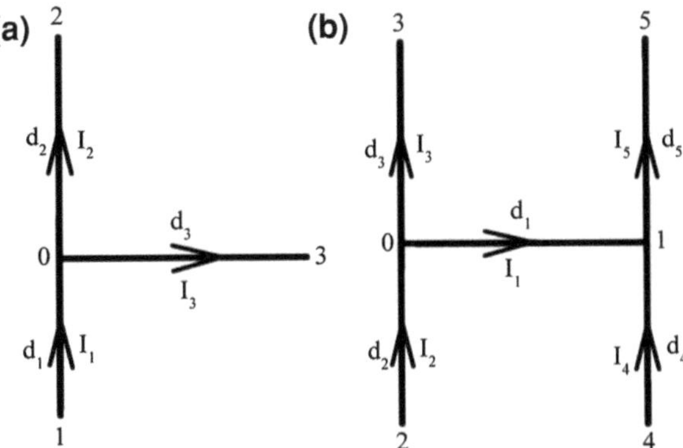

Fig. 5.1 The schematic diagrams for (**a**) T-shape and (**b**) H-shape metallic wire structures. The current directions, wire lengths, wire ends, and joints are indicated (reprinted with permission from Ref. [4]; copyright (2009), American Institute of Physics)

of resonance modes is generally dictated by the symmetry of a network. The eigenvectors suggest that resonance modes originate from charge conservation and energy conservation, the current flow acts as a "kinetic energy" while the boundary electric fields at ends and joints act as a "potential energy", the oscillation between the two energies defines the various eigenmodes in the wire network. Furthermore, the electric and the magnetic responses are clearly indicated by the charge and the current patterns within a wire network.

For the T-shape wire network as shown in Fig. 5.1a, the current density $\vec{J}(\vec{r})$ can be written in terms of the wire currents $I_i(\vec{r})$ ($i = 1, 2, 3$) as,

$$\vec{J}(\vec{r}) = I_1(y)\delta(x)\delta(z)\hat{e}_y + I_2(y)\delta(x)\delta(z)\hat{e}_y + I_3(x)\delta(y)\delta(z)\hat{e}_x,$$

$\hat{e}_x$ and $\hat{e}_y$ are unit vectors along x and y axes, $\delta(x)$ is Dirac's delta function. Substituting this equation into Eq. (5.2a, 5.2b), we arrive at the equations

$$\vec{E}_L(\vec{r}) = -\frac{i\omega}{c^2}\hat{e}_y \int_{-d_1}^{0} \frac{I_1(y')}{|\vec{r} - y'\hat{e}_y|}\, dy' - \frac{i\omega}{c^2}\hat{e}_y \int_{0}^{d_2} \frac{I_2(y')}{|\vec{r} - y'\hat{e}_y|}\, dy' - \frac{i\omega}{c^2}\hat{e}_x \int_{0}^{d_3} \frac{I_3(x')}{|\vec{r} - x'\hat{e}_x|}\, dx',$$

$$(5.13)$$

$$\vec{E}_C(\vec{r}) = +\frac{1}{i\omega}\vec{\nabla} \int_{-d_1}^{0} \frac{I_1'(y')}{|\vec{r} - y'\hat{e}_y|}\, dy' + \frac{1}{i\omega}\vec{\nabla} \int_{0}^{d_2} \frac{I_2'(y')}{|\vec{r} - y'\hat{e}_y|}\, dy'$$

$$+\frac{1}{i\omega}\vec{\nabla} \int_{0}^{d_3} \frac{I_3'(x')}{|\vec{r} - x'\hat{e}_x|}\, dx'.$$

$$(5.14)$$

d_i is the wire length and $I_i'(x) = dI_i(x)/dx$ is the derivative of the current along wire i. Within the equivalent circuit theory, one seeks the relationship between the wire currents and the tangential components of electric field along the wires. The wire radius a_i comes into play since the tangential field is taken on the wire surface. To extract the inductances and capacitances of the wire network, we expand both the electric fields and the currents with a complete set of orthogonal functions. For the problem under consideration, the plane wave expansion is a good choice for each segment of the metallic wire.

It is straightforward to write down the integrals for the impedance matrix elements. For example,

$$L^{21}(h, h') = \int\limits_{0}^{d_2} \int\limits_{-d_2}^{0} \exp[i\pi(-hy/d_2 + h'y'/d_1)]/[(y - y')^2 + a_2^2]^{1/2} dy\, dy'$$

All nine matrix elements of the reduced inductance are listed in [4]. We have used an approximation wherein the radius a provides a cutoff in the denominator. It is simple to check that $L^{ij}(h_i, h'_j)$ is a Hermitian matrix if the wire thicknesses are all taken to be the same. In the absence of resistances and external fields, the eigenvalue Eq. (5.3) has the Hermitian property, thus dissipationless modes are guaranteed.

For a finite wire network with wire ends and joints, the circuit Eq. (5.5) have to be supplemented by boundary conditions. For the T-shape network, there are three wire ends and one joint, the four boundary conditions are

$$\sum_{h_i} (-1)^{h_i} I_i(h_i) = 0, \quad \text{for } i = 1, 2, 3$$

and

$$-\sum_{h_1} I_1(h_1) + \sum_{h_2} I_2(h_2) + \sum_{h_3} I_3(h_3) = 0. \tag{5.15}$$

As is discussed above, these conditions can be summarized in the form of Eq. (5.8) as $\boldsymbol{H_0 I} = 0$. For the present case for $i = 1, 2, 3$

$$H_0(i; jh) = \delta_{i,j}(-1)^h$$

$$H_0(4; 1h) = -1, \quad H_0(4, 2h) = 1, \quad H_0(4, 3h) = 1.$$

To enforce the boundary conditions we introduce the boundary fields at ends of the wire segments. Let us denote the magnitude of the boundary field at the ith wire end by α_i and the one at the origin by α_0, the final circuit equations can be written as

$$\frac{2cd_1^2}{i\pi\Omega_1} E_{\text{ext}}(h_1) = \frac{2cd_1 R_1}{i\pi\Omega_1} I_1(h_1) + \sum_{j, h'_j} \left[\hat{e}_1 \cdot \hat{e}_j - \frac{h_1 h'_j}{\Omega_1 \Omega_j} \right] L^{1j}(h_1, h'_j) I_j(h'_j) - \frac{(-1)^{h_1}}{\Omega_1} \sigma_1$$

$$+ \frac{1}{\Omega_1} \sigma_0,$$

$$\frac{2cd_2^2}{i\pi\Omega_2}E_{\text{ext}}(h_2) = \frac{2cd_2R_2}{i\pi\Omega_2}I_2(h_2) + \sum_{j,h_j'}\left[\hat{e}_2\cdot\hat{e}_j - \frac{h_2h_j'}{\Omega_2\Omega_j}\right]L^{2j}(h_2,h_j')I_j(h_j') + \frac{(-1)^{h_2}}{\Omega_1}\sigma_2 - \frac{1}{\Omega_1}\sigma_0,$$

$$\frac{2cd_3^2}{i\pi\Omega_3}E_{\text{ext}}(h_3) = \frac{2cd_3R_3}{i\pi\Omega_3}I_3(h_3) + \sum_{j,h_j'}\left[\hat{e}_3\cdot\hat{e}_j - \frac{h_3h_j'}{\Omega_3\Omega_j}\right]L^{3j}(h_3,h_j')I_j(h_j') + \frac{(-1)^{h_3}}{\Omega_1}\sigma_3$$
$$- \frac{1}{\Omega_1}\sigma_0,$$

$$(5.16)$$

$\Omega_i = \omega d_i/(\pi c)$, $\sigma_i = (cd_1/i\pi)\alpha_i$ is the reduced magnitude of the boundary field. Equation (5.16) can also be summarized as (for $k = 1, 2, 3$)

$$E_{\text{ext}}'(h_k) = \sum_{j,h_j'}X_{kj}(h_k,h_j')I_j(h_j') + (-1)^{s_k}\left[-\frac{(-1)^{h_k}}{\omega}\sigma_k' + \frac{\sigma_0'}{\omega}\right]. \qquad (5.16a)$$

where $E_{\text{ext}}'(h_k) = \frac{2c^2d_k}{i\omega}E_{\text{ext}}(h_k)$, $\sigma_i' = c\pi\sigma_i/d_1$, $s_1 = 0$, $s_{2,3} = 1$,

$$X_{kj}(h_k,h_j') = \frac{2c^2R_k}{i\omega}\delta_{k,j}\delta(h_j' - h_k) + \left[\hat{e}_k\cdot\hat{e}_j - \frac{h_kh_j'^2c^2}{d_kd_j\omega^2}\right]L^{kj}(h_k,h_j'). \qquad (5.16b)$$

To determine a resonance, we determine nontrivial solutions when $E_{\text{ext}} = 0$ and get

$$I_j(h_j') = \sum_{k,h_k}X_{jk}^{-1}(h_j',\,h_k)(-1)^{s_k}\left[-\frac{(-1)^{h_k}}{\omega}\sigma_k' + \frac{\sigma_0'}{\omega}\right].$$

where the inverse of X is defined by $\sum_{j,h_j}X_{kj}(h_k,h_j)X_{jl}^{-1}(h_j,h_l) = \delta_{kl}\delta(h_k - h_l)$.

Now we substitute this back into the boundary conditions (5.15) and get the 4×4 matrix Eq. (5.10b):

$$H_0Y^{-1}\sigma = 0. \qquad (5.17a)$$

These can be written explicitly as

$$\sum_{h_j'}(-1)^{h_j'}\sum_{k,h_k}X_{jk}^{-1}(h_j',h_k)[-(-1)^{h_k}\sigma_k' + \sigma_0'](-1)^{s_k} = 0, \quad \text{for } j = 1, 2, 3$$

and

$$-\sum_{j,h_j'}(-1)^{s_j}\sum_{k,h_k}X_{jk}^{-1}(h_j',h_k)[-(-1)^{h_k}\sigma_k' + \sigma_0'](-1)^{s_k} = 0. \qquad (5.17b)$$

We discuss the solutions of Eq. (5.17b) next.

5.3.1 Classifying the Eigenmodes by Network Symmetry

The equation set (5.17b) is somewhat different from the usual eigenvalue equations since the resonance frequency appears nonlinearly as a parameter of the operator. As is discussed above, to obtain the resonance frequency we solve the pseudo-eigenvalue equations

$$I_i(x) = \mathrm{Re} \sum_{h_i} I_i(h_i) \exp\left(\frac{i\pi h_i x}{d_i}\right),$$

$$Q_i(x) = -\frac{1}{\Omega_1 c}\frac{d_1}{d_i}\mathrm{Im} \sum_{h_i} h_i I_i(h_i) \exp\left(\frac{i\pi h_i x}{d_i}\right). \tag{5.18}$$

$$\boldsymbol{H_0 Y^{-1} \sigma = \lambda \sigma}$$

for the pseudo-eigenvalues λ for a distribution of frequencies ω. Because the determinant is equal to the product of the eigenvalues Eq. (5.17a) is satisfied if one of the eigenvalues λ is "zero". Because we have included the electrical resistances in our calculation, these pseudo-eigenvalues are not real but their imaginary parts are usually small. The condition of a zero λ is determined as follows. The number of the pseudo-eigenvalues λ with positive real-part is counted at each frequency ω, a resonance usually occurs when there is a change in this number. In our calculation, we found that some pseudo-eigenvalues change sign but does not cross zero as the frequency is changed. We have made sure that the frequency mesh size is small enough so that there is no confusion in picking the correct eigenvector.

To facilitate the numerical calculation, the metal wires are assumed to be made of copper with resistivity $\rho = 1.8 \times 10^{-5}\ \Omega$ mm at room temperature. All coordinates are made dimensionless and positive by defining a dimensionless length $u = (|x_i|, |y_i|)/d_i$, $u = 0$ and $u = 1$ correspond to the wire joints and ends. The dimensionless frequency $\Omega = \Omega_1 = \omega d_1/\pi c$ is always defined with respect to d_1. The wires are treated as uniform and for illustrative purposes, the radii of the wires are set as $a_i = 0.1$ mm. Throughout this paper, a cutoff in wave vector corresponding to $h_{\max} = 13$ is adopted to ensure that the eigenfrequencies, the current, and the charge profiles are all convergent to three effective digits.

We discuss the current and the charge profiles of various eigenmodes as the structure symmetry varies. We first consider the case with the highest symmetry, $d_1 = d_2 = d_3 = 10$ mm. Since all three channels ($d_1 + d_2$, $d_2 + d_3$, and $d_3 + d_1$) have the same length, one expects for the lowest frequency mode there is equal probability of finding currents of the same magnitude in each channel. However, the mirror symmetry with respect to wire 3 suggests that even and odd symmetrical solutions exist between currents in wires 1 and 2. Three types of eigenmodes are possible: (1) $I_1(u) = I_2(u)$ and $I_3(u) \equiv 0$; (2) $I_1(u) = -I_2(u) = 0.5 I_3(u)$; and (3) the three currents are zero at the ends of the respective wires and are not related by the current conservation condition. Indeed, we find the three representative

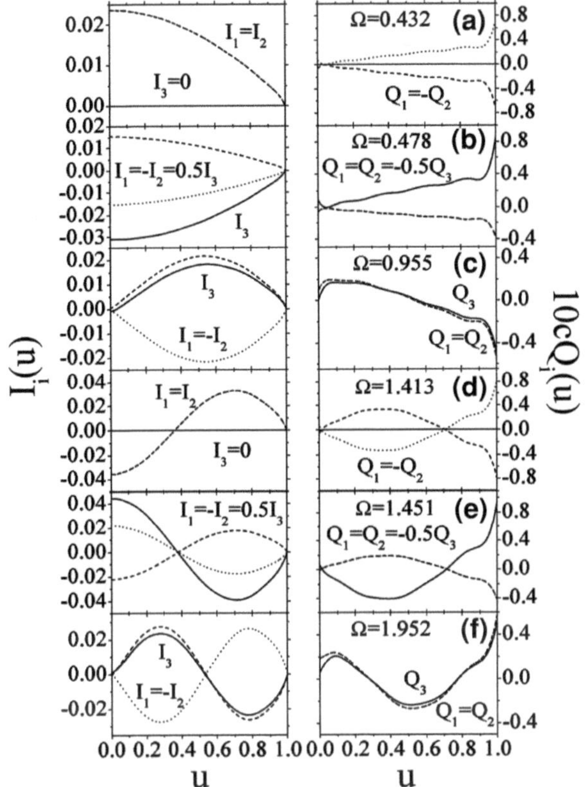

Fig. 5.2 The current and charge distributions for the eigenmodes in a T-shape wire structure with $d_1 = d_2 = d_3 = 10$ mm. *Left panels* are for the currents and *right panels*, for the charge distributions. Wires 1–3 are denoted by *dotted*, *dashed*, and *solid lines*, respectively. The wire joint (0) and wire ends (1, 2, 3) correspond to $u = 0$ and $u = 1$. The symmetries among current and charge components are indicated (reprinted with permission from Ref. [4]; copyright (2009), American Institute of Physics)

examples presented in Fig. 5.2a–c for the lowest frequency modes with $\Omega = 0.432$, 0.478, and 0.955 (see also Table 5.1). The resonance wavelength for the lowest mode, $4.6d_1$, is close to twice the total path length, $4d_1$, that one expects for a thin straight wire with no joints.

The boundary electric fields σ_i are listed in Table 5.1, the positive field drives the charge away from the boundaries while the negative field draws the charge back to wire ends. The symmetry of the currents is also reflected in the fields σ_i. When the current is zero, the local electric fields σ_i are zero at the ends. A similar phenomenon is also found between the current and the charge distributions.

The charge densities are also shown in Fig. 5.2. To plot the charge distribution on the same scale, we multiply $Q(u)$ by $10c$ with c the light velocity in vacuum. There are charges accumulated at wire ends and the boundary fields are partly the

Table 5.1 The dimensionless eigenfrequency Ω and the boundary fields σ at wire ends and joints for the eigenmodes in three different T-shape structures labeled in the left column

	$I_i(u),Q_i(u)$	Ω	σ_0	σ_1	σ_2	σ_3
	Fig. 5.2a	0.432	0.000	0.704	−0.704	0.000
$d_1 = 10$ mm	Fig. 5.2b	0.478	−0.038	−0.377	−0.377	0.839
$d_2 = 10$ mm	Fig. 5.2c	0.955	0.545	−0.495	−0.495	−0.432
$d_3 = 10$ mm	Fig. 5.2d	1.413	0.000	0.683	−0.683	0.000
	Fig. 5.2e	1.451	0.025	−0.372	−0.372	0.813
	Fig. 5.2f	1.952	0.543	0.485	0.485	0.429
	Fig. 5.3a	0.287	−0.277	−0.414	−0.414	0.759
$d_1 = 10$ mm	Fig. 5.3b	0.432	0.000	−0.704	0.704	0.000
$d_2 = 10$ mm	Fig. 5.3c	0.683	−0.260	0.416	0.416	0.754
$d_3 = 20$ mm	Fig. 5.3d	0.973	0.540	−0.495	−0.495	0.443
	Fig. 5.3e	1.268	−0.260	0.392	0.392	0.759
	Fig. 5.3f	1.413	0.000	0.683	−0.683	0.000
	Fig. 5.4a	0.187	−0.218	−0.206	−0.602	0.739
$d_1 = 10$ mm	Fig. 5.4b	0.299	0.313	0.684	−0.590	−0.285
$d_2 = 20$ mm	Fig. 5.4c	0.484	−0.040	0.666	0.079	0.734
$d_3 = 30$ mm	Fig. 5.4d	0.637	−0.361	0.563	0.669	−0.302
	Fig. 5.4e	0.792	0.225	−0.278	0.539	0.750
	Fig. 5.4f	0.979	0.540	−0.495	0.492	−0.444

results of such charge distributions. The higher frequency eigenmodes correspond to current distributions with additional nodes and are shown in Fig. 5.2d–f. The symmetries of these modes are not altered, the eigenfrequency has increased to $\Omega = 1.413$, 1.451,, and 1.952. From the charge distributions, we obtain finite electric dipole moments along the y axis for modes (a, d) and along the x axis for modes (b, c, e, f). We next look at the magnetic moment.

For open wire structures with a net total current $\vec{I}_t = \sum_i \int d^3r \vec{J}_i(r)$ that is not equal to zero, there is a parallel axis theorem for the value of the magnetic dipole moment M so that when the origin of the coordinate system is moved by a distance $\boldsymbol{D}$, the change in the magnetic dipole moment is equal to $\boldsymbol{I}_t \times \boldsymbol{D}/2c$. For the T structure, if we take the origin to be at 0 in Fig. 5.1a, then all magnetic dipole moments are zero.

To change the symmetry among the three channels while keeping the mirror symmetry intact, we consider an example with $d_3 = 2d_1 = 2d_2 = 20$ mm. Since the two longest channels (30 mm) all involve wire 3, the ordering of eigenmodes is completely rearranged. As seen from the lowest six eigenmodes presented in Fig. 5.3a–f, the eigenfrequencies are significantly reduced in comparison with those in Fig. 5.2 due to a longer wire 3. The resonance wavelength $6.9d_1$ is not far away from $6d_1$, twice the longest path length. Regardless of the wire length, the current modulation wavelengths are all compatible in the three wires. Figure 5.3 suggests that the second mode is the same as the first mode in the previous case except that now I_1 is no longer equal to $0.5I_3$. The electric polarization is along y-axis for modes (b, f) and along x-axis for the rest of the eigenmodes.

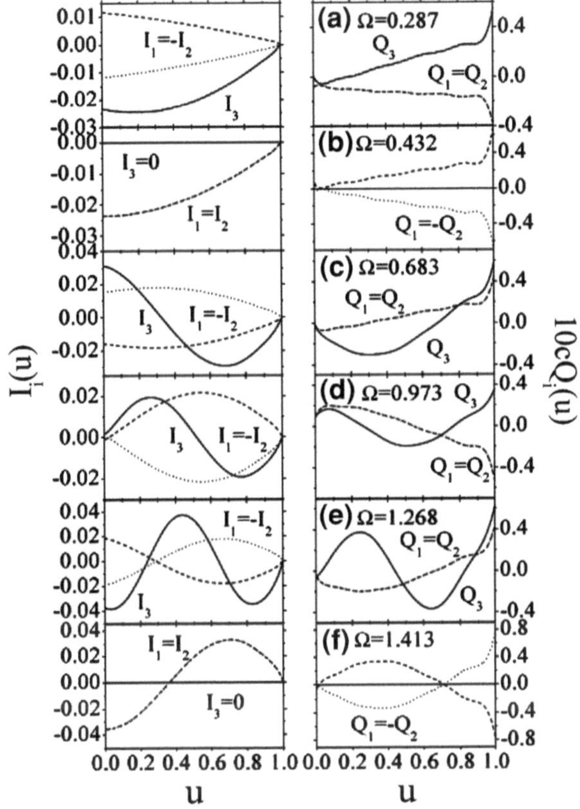

Fig. 5.3 The current and charge distributions for the eigenmodes in a T-shape wire structure with $2d_1 = 2d_2 = d_3 = 20$ mm. The other notations are the same as Fig. 5.2 (reprinted with permission from Ref. [4]; copyright (2009), American Institute of Physics)

We also consider a general case $2d_3 = 3d_2 = 6d_1 = 60$ mm with no symmetry. The corresponding current and charge profiles are shown in Fig. 5.4a–f for the low frequency eigenmodes. As no more constraint is imposed on I_1 and I_2, this leads to the absence of first two types of solutions. The main feature of the eigenmodes is the active involvement of wire 3 in all low frequency modes. The lowest eigenmode shown in Fig. 5.4a involves the longest wire segments 2 and 3. The resonance wavelength, $10.69d_1$, is close to $10d_1$, twice the longest path length. Also the three wires are generally coupled in each eigenmode and the response property is more sensitive to specific dimensions of the wire network. The electric response property is similar to those discussed above.

5.3.2 Physical Picture

The physics behind the nature of the symmetry of the solutions is captured in the simplest approximation to the solution of the circuit equation where we ignore the mutual inductance and capacitance and keep only the diagonal terms in the Fourier transform of the circuit parameters. We describe this next.

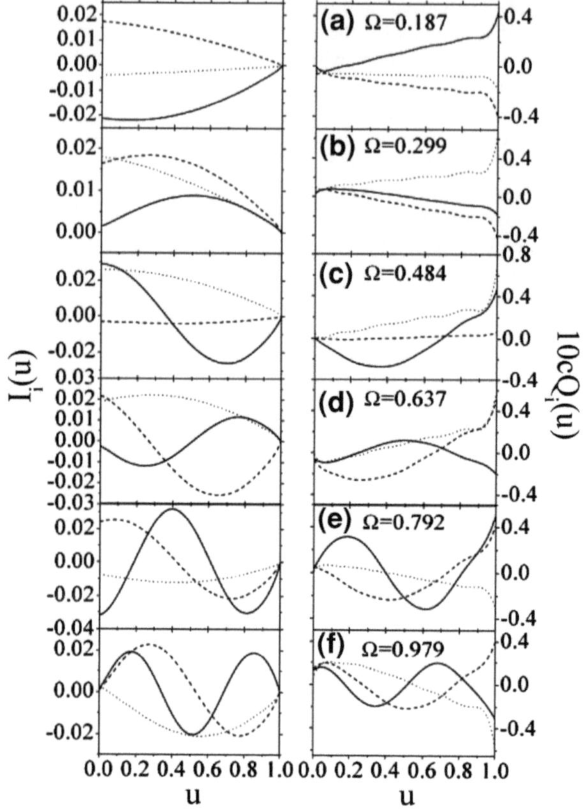

Fig. 5.4 The current and charge distributions for the eigenmodes in a T-shape wire structure with $6d_1 = 3d_2 = 2d_3 = 60$ mm. The other notations are the same as Fig. 5.2 (reprinted with permission from Ref. [4]; copyright (2009), American Institute of Physics)

Retaining only the log divergent diagonal components of the Fourier transform and ignoring the mutual capacitances and inductances, we obtain

$$X_{ab}(h, h') \approx x_{aa}(h)\delta_{a,b}\delta(h - h').$$

$$X_{ab}^{-1}(h, h') \approx x_{aa}^{-1}(h)\delta_{ab}\delta(h - h').$$

Equation (5.17b) can be written as

$$y(j)\sigma_j' = z(j)\sigma_0', \quad \text{for } j = 1, 2, 3 \tag{5.19a}$$

and

$$\sigma_0' = \sum_k z(k)\sigma_k'/W = t, \tag{5.19b}$$

for the condition that the sum of the three currents going into vertex zero is conserved. Here $y(j) = \sum_h x_{jj}^{-1}(h)$, $z(j) = \sum_h x_{jj}^{-1}(h)(-1)^h$, $t = \sum_k z(k)\sigma_k'/W$,

$$W = \sum_j y(j).$$

If all three wires are the same, Eq. (5.19a) becomes

$$y\sigma_j' = z\sigma_0',$$

This equation is satisfied if (a) $\sigma_0' = 0$ and $y = 0$ or (b) $\sigma_j' = z\sigma_0'/y$. The other equation is

$$y\sigma_0' = z\sum_k \sigma_k'/3,$$

For case (a), so long as $\sum_k \sigma_k' = 0$, this equation is satisfied. This corresponds to the resonances 1, 2, 4, 5 discussed above.

For case (b), we get $z = \pm Y$. Thus $\sigma_{1,2,3} = \pm\sigma_0$. This corresponds to the resonances 3 (for the $-$ sign) and 6 (for the $+$ sign).

5.4 The H Structure

Although the T-shape network is very informative in illustrating the relationship among the current, the charge distributions in the wires and the boundary fields at the wire ends and joints, its response is mainly electric and does not shed significant insight on the material design with negative refractive index properties, which requires a significant magnetic response. We have also applied our formalism to the H-shape structures which can exhibit comparable magnetic and electric responses. Rich patterns of eigenmodes emerge because of the more complex wire configuration.

For the H-shape metallic wire network as shown in Fig. 5.1b, the derivation of the circuit equations is straightforward, thus only the final circuit equations are presented below:

$$\frac{2cd_1^2}{i\pi\Omega_1}E_{\text{ext}}(h_1) = \frac{2cd_1R_1}{i\pi\Omega_1}I_1(h_1) + \sum_{j,h_j'}\left[\hat{e}_1 \cdot \hat{e}_j - \frac{h_1 h_j'}{\Omega_1\Omega_j}\right]L^{1j}(h_1, h_j')I_j(h_j') + \frac{(-1)^{h_1}}{\Omega_1}\sigma_1$$
$$- \frac{1}{\Omega_1}\sigma_0,$$

$$\frac{2cd_2^2}{i\pi\Omega_2}E_{\text{ext}}(h_2) = \frac{2cd_2R_2}{i\pi\Omega_2}I_2(h_2) + \sum_{j,h_j'}\left[\hat{e}_2 \cdot \hat{e}_j - \frac{h_2 h_j'}{\Omega_2\Omega_j}\right]L^{2j}(h_2, h_j')I_j(h_j') - \frac{(-1)^{h_2}}{\Omega_1}\sigma_2$$
$$+ \frac{1}{\Omega_1}\sigma_0,$$

$$\frac{2cd_3^2}{i\pi\Omega_3}E_{\text{ext}}(h_3) = \frac{2cd_3R_3}{i\pi\Omega_3}I_3(h_3) + \sum_{j,h_j'}\left[\hat{e}_3\cdot\hat{e}_j - \frac{h_3h_j'}{\Omega_3\Omega_j}\right]L^{3j}(h_3,h_j')I_j(h_j') + \frac{(-1)^{h_3}}{\Omega_1}\sigma_3$$
$$- \frac{1}{\Omega_1}\sigma_0,$$

$$\frac{2cd_4^2}{i\pi\Omega_4}E_{\text{ext}}(h_4) = \frac{2cd_4R_4}{i\pi\Omega_4}I_4(h_4) + \sum_{j,h_j'}\left[\hat{e}_4\cdot\hat{e}_j - \frac{h_4h_j'}{\Omega_4\Omega_j}\right]L^{4j}(h_4,h_j')I_j(h_j') - \frac{(-1)^{h_4}}{\Omega_1}\sigma_4$$
$$+ \frac{1}{\Omega_1}\sigma_1,$$

$$\frac{2cd_5^2}{i\pi\Omega_5}E_{\text{ext}}(h_5) = \frac{2cd_5R_5}{i\pi\Omega_5}I_5(h_5) + \sum_{j,h_j'}\left[\hat{e}_5\cdot\hat{e}_j - \frac{h_5h_j'}{\Omega_5\Omega_j}\right]L^{5j}(h_5,h_j')I_j(h_j') + \frac{(-1)^{h_5}}{\Omega_1}\sigma_5$$
$$- \frac{1}{\Omega_1}\sigma_1.$$

$$(5.20)$$

All the notations are exactly the same as in T-shape case except that now we have five metallic wires forming the network. There are also four wire ends ($i = 2, 3, 4, 5$) and two joints ($i = 0, 1$), so that six boundary fields σ_i are introduced to implement the six boundary conditions in the network. The reduced inductance in the H-shape network is a 5×5 matrix which is listed in [4]. For example,

$$L^{41}(h,h') = \int_{-d_4}^{0}\int_{0}^{d_1} \exp[i\pi(-hy/d_4 + h'x'/d_1)]/[y^2 + (x' - d_1)^2 + a_4^2]^{1/2}.$$

The above equations can be written in a more compact form similar to Eq. (5.16a) as

$$E_{\text{ext}}'(h_k) = \sum_{j,h_j'}X_{kj}(h_k,h_j')I_j(h_j') + (-1)^{s_k}\left[-\frac{(-1)^{h_k}}{\omega}\sigma_k' + \frac{\sigma_0'}{\omega}\right], \quad \text{for } k = 1,2,3,$$

$$E_{\text{ext}}'(h_k) = \sum_{j,h_j'}X_{kj}(h_k,h_j')I_j(h_j') + (-1)^{s_k}\left[-\frac{(-1)^{h_k}}{\omega}\sigma_k' + \frac{\sigma_1'}{\omega}\right], \quad \text{for } k = 4,5.$$

$$(5.20a)$$

Here $s_{1,3,5} = 1, s_{2,4} = 0$, X is the same as is given in Eq. (5.16b).

The boundary conditions are:

$$\sum_{h_i}(-1)^{h_i}I_i(h_i) = 0, \quad \text{for } i = 2\text{--}5$$

$$\sum_{h_1} I_1(h_1) - \sum_{h_2} I_2(h_2) + \sum_{h_3} I_3(h_3) = 0.$$

$$\sum_{h_1} (-1)^{h_1} I_1(h_1) - \sum_{h_5} I_5(h_5) + \sum_{h_4} I_4(h_4) = 0. \tag{5.21}$$

Again these conditions can be summarized in the form of Eq. (5.8) as $H_0 I = 0$. For the present case with

$$H_0(i; jh) = \delta_{i,j}(-1)^h, \quad \text{for } i = 2, 3, 4, 5,$$

$$H_0(1; 1h) = -1, \quad H_0(1, 2h) = -1, \quad H_0(1, 3h) = 1,$$

$$H_0(6; 1h) = (-1)^h, \quad H_0(6, 4h) = 1, \quad H_0(6, 5h) = -1.$$

We first consider the simplest and most symmetric situation with $d_1 = d_2 = d_3 = d_4 = d_5 = 10\,\text{mm}$. This configuration has both perpendicular and horizontal mirror reflection symmetries, thus we expect eigenmodes that satisfy the relationships $I_2(u) = \pm I_3(u) = \pm I_4(u) = \pm I_5(u)$. (The left and right joints of the wire 1 in the H-shape structure are denoted by $u = 0$ and $u = 1$, respectively.) As the longest channels are the $(2, 3 \leftrightarrow 4, 5)$ crossing paths, the resonance configurations with nonzero currents along these channels form the lowest eigenmode.

Figure 5.5a shows the lowest eigenmode at $\Omega = 0.275$. One can think of an H as formed from two Ts. The mode in Fig. 5.5a can also be thought of as formed by superimposing the modes in Fig. 5.1a for the T structure. The corresponding boundary electric fields are listed in Table 5.2. For this mode the upper and lower circuits form opposite magnetic dipoles so the whole circuit behaves like a magnetic quadrupole, the dielectric response resembles a dipole along x-axis. Figure 5.5b, c is two decoupled eigenmodes where the current along the horizontal segment 1 are zero. Just as for the T structure, the local fields at the ends of segments 1, σ_0, σ_1 are zero. Vertical currents along 2–3 and 4–5 flow either parallel or antiparallel to each other. The parallel current eigenmode acts as an electric dipole along the y-axis while the antiparallel eigen mode acts as a magnetic dipole and an electric quadrupole. The total current I_t for this mode is zero and the magnetic dipole moment does not depend on the choice of the origin of the coordinates. The same type of mode is seen in two parallel wires. This type of mode serves as a candidate for exploring high frequency magnetism and has been exploited in the fishnet structure.

For the eigenmode in Fig. 5.5d there is a node in the current along the horizontal wire 1, the current pattern indicates that four magnetic dipoles are formed at the four triangular corners, thus the circuit behaves as a magnetic hexadecapole, this mode is like a dark mode and may be used as a gain media. The electric response is of quadrupole type. Figure 5.5e corresponds to a somewhat decoupled mode in that the magnitude of the currents at the joints is very small. Its electromagnetic (EM) property is similar to the lowest eigenmode in Fig. 5.5a. The current pattern for the high frequency mode in Fig. 5.5f has a node in every wire, its property is similar to the eigenmode shown in Fig. 5.5d.

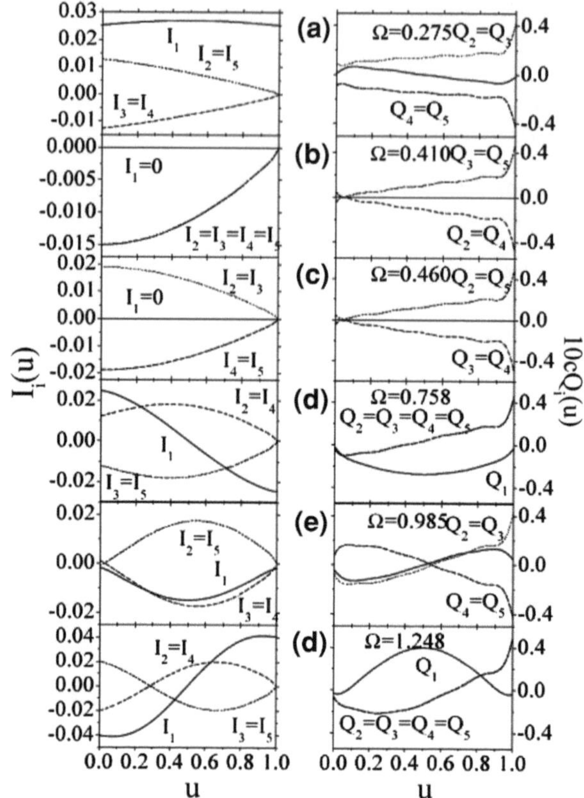

Fig. 5.5 The current and charge distributions for the eigenmodes in an H-shape wire structure with $d_1 = d_2 = d_3 = d_4 = d_5 = 10$ mm. The wires 1–5 are denoted by *solid, thick dotted, dotted, thick dashed,* and *dashed lines,* respectively. $u = 0$ and $u = 1$ correspond to the positions 0 and 1 of Fig. 5.1b for wire 1, 0 and (2, 3) for wires 2 and 3, and 1 and (4, 5) for wires 4 and 5, respectively. The symmetries among current and charge components are as indicated (reprinted with permission from Ref. [4]; copyright (2009), American Institute of Physics)

There are several ways of lowering the symmetry of the H-shape structures. As the H-shape structure with equal wire lengths has both perpendicular and horizontal mirror symmetry planes, one can remove either the perpendicular or the horizontal mirror symmetry or both, these three configurations are discussed separately below.

For the case with $d_1 = d_2 = d_3 = 2d_4 = 2d_5 = 20$ mm, the horizontal mirror symmetry guarantees that $I_2(u) = \pm I_3(u)$ and $I_4(u) = \pm I_5(u)$. The eigenmodes can be viewed as basically evolving from those of Fig. 5.5 although the huge "local" resonance frequency mismatch between the (d_2, d_3) branch and (d_4, d_5) branch can make the current strength $I_{2,3}$ very much different from $I_{4,5}$ when these two branches are nearly decoupled.

The lowest eigenmode in Fig. 5.6a is quite similar to Fig. 5.5a, since the sum of the currents is connected via the cross wire 1, the magnitude of the currents are

Table 5.2 The resonance frequencies and the boundary fields at wire ends and joints for the eigenmodes in four H-shape structures of different symmetries

	$I_i(u),Q_i(u)$	Ω	σ_0	σ_1	σ_2	σ_3	σ_4	σ_5
	Fig. 5.5a	0.275	0.311	−0.311	0.448	0.448	−0.448	−0.448
$d_1 = 10$ mm	Fig. 5.5b	0.410	0.000	0.000	−0.498	0.498	−0.498	0.498
$d_2 = 10$ mm	Fig. 5.5c	0.460	0.000	0.000	0.498	−0.498	−0.498	0.498
$d_3 = 10$ mm	Fig. 5.5d	0.758	−0.352	−0.352	0.429	0.429	0.429	0.429
$d_4 = 10$ mm	Fig. 5.5e	0.985	−0.430	0.430	0.390	0.390	−0.390	−0.390
$d_5 = 10$ mm	Fig. 5.5f	1.248	−0.308	−0.308	0.440	0.440	0.440	0.440
	Fig. 5.6a	0.332	−0.194	0.451	−0.355	−0.355	0.503	0.503
$d_1 = 20$ mm	Fig. 5.6b	0.442	0.000	0.000	0.704	−0.704	0.066	−0.066
$d_2 = 20$ mm	Fig. 5.6c	0.844	−0.511	0.072	0.528	0.528	0.295	0.295
$d_3 = 20$ mm	Fig. 5.6d	0.866	0.000	0.000	0.011	−0.011	−0.706	0.706
$d_4 = 10$ mm	Fig. 5.6e	1.176	−0.452	0.104	0.509	0.509	−0.363	−0.363
$d_5 = 10$ mm	Fig. 5.6f	1.431	0.000	0.000	−0.705	0.705	0.018	−0.018
	Fig. 5.7a	0.309	−0.314	0.314	−0.541	−0.329	0.541	0.329
$d_1 = 20$ mm	Fig. 5.7b	0.535	−0.187	−0.187	0.535	−0.422	0.535	−0.422
$d_2 = 20$ mm	Fig. 5.7c	0.649	−0.243	0.243	0.380	−0.544	−0.380	0.544
$d_3 = 10$ mm	Fig. 5.7d	0.993	−0.086	−0.086	0.130	0.689	0.130	0.689
$d_4 = 20$ mm	Fig. 5.7e	1.252	0.274	−0.274	−0.454	−0.466	0.454	0.466
$d_5 = 10$ mm	Fig. 5.7f	1.333	−0.223	−0.223	0.612	0.271	0.612	0.271
	Fig. 5.8a	0.305	0.304	−0.304	0.564	0.298	−0.298	−0.564
$d_1 = 20$ mm	Fig. 5.8b	0.581	−0.205	−0.205	0.541	−0.405	−0.405	0.541
$d_2 = 20$ mm	Fig. 5.8c	0.601	0.236	−0.236	−0.372	0.553	−0.553	0.372
$d_3 = 10$ mm	Fig. 5.8d	1.003	−0.070	−0.070	0.114	0.693	0.693	0.114
$d_4 = 10$ mm	Fig. 5.8e	1.252	0.274	−0.274	−0.455	−0.465	0.465	0.455
$d_5 = 20$ mm	Fig. 5.8f	1.335	−0.228	−0.228	0.612	0.266	0.266	0.612

roughly the same. The resonance corresponds to an electric dipole and a magnetic quadrupole. The uncoupled modes shown in Figs. 5.6b, d are derived from Fig. 5.5b (parallel mode) and Fig. 5.5c (antiparallel mode) . However, the local resonance frequency mismatch seriously enhances the magnitude disparity between the long and the short wires. More precisely, the resonance wavelength for mode b (d) is close to the sum of the lengths of wire 2 and 3 (4 and 5). The absolute magnitude of the currents in wires 4 and 5 (2 and 3) become much smaller. Both the parallel and the antiparallel modes behave as an electric dipole along the y-axis; the antiparallel mode also possesses a certain amount of magnetic dipole character. Similarly, the eigenmodes shown in Fig. 5.6c, e can be traced from Fig. 5.5d, f. Figure 5.6f corresponds to a high frequency mode not included in Fig. 5.5, it has purely dipole-like electric response.

By taking $d_1 = d_2 = 2d_3 = d_4 = 2d_5 = 20$ mm we can also remove the horizontal mirror symmetry plane. One of the outcome of the broken horizontal mirror symmetry is the absence of $I_1(u) \equiv 0$ solutions, the decoupled solutions in Fig. 5.5 transform into coupled modes. However, the perpendicular mirror symmetry still imposes the constraints that $I_2(u) = \pm I_4(u)$ and $I_3(u) = \pm I_5(u)$. Taking these factors into consideration, the eigenmodes can still be traced back from those

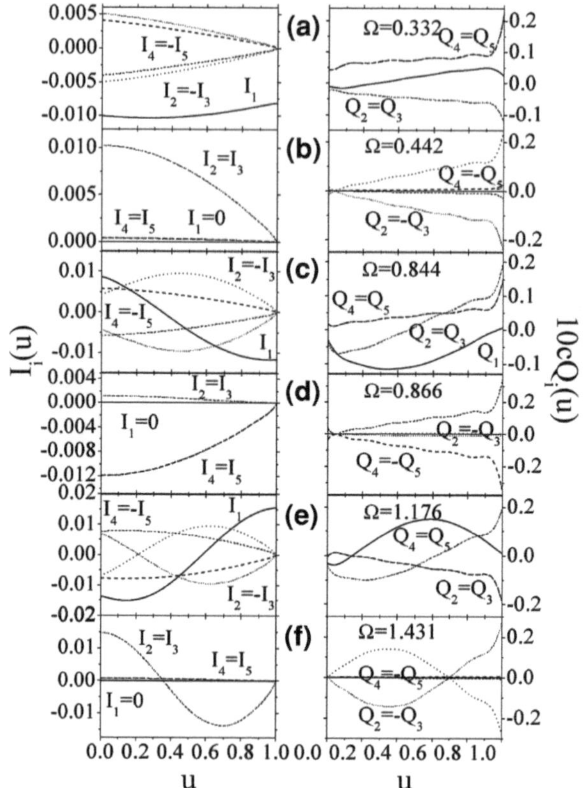

Fig. 5.6 The current and charge distributions for the eigenmodes in a H-shape wire structure with $d_1 = d_2 = d_3 = 2d_4 = 2d_5 = 20$ mm. The other notations are the same as Fig. 5.5 (reprinted with permission from Ref. [4]; copyright (2009), American Institute of Physics)

presented in Fig. 5.5a–f. For example, the mode shown in Fig. 5.7a has an electric dipole along x-axis and a magnetic dipole; the response of mode (b) behaves as an electric dipole along the y-axis and a magnetic quadrupole; the mode (c) can be modeled by an electric quadrupole and a magnetic dipole while mode (d) corresponds to an electric quadrupole and magnetic quadrupole. Similarly, modes (e) and (f) have the same symmetry as modes (c) and (d) except for the extra node in the current pattern, corresponding to the higher resonance frequency. Because of the absence of the horizontal symmetry, magnetic multipoles from the upper and lower circuits do not cancel exactly, the higher order electric and magnetic multipoles of Fig. 5.5 are usually reduced to lower order multipoles. Thus, the number of eigenmodes with nonzero magnetic dipole increases, which might offer more options for designing circuits with the so-called high frequency magnetism.

The last case we consider in this chapter is the case with neither horizontal nor vertical mirror symmetry, but with a space inversion symmetry. This is realized by taking $d_1 = d_2 = 2d_3 = 2d_4 = d_5 = 20$ mm. The lowest six eigenmodes are shown

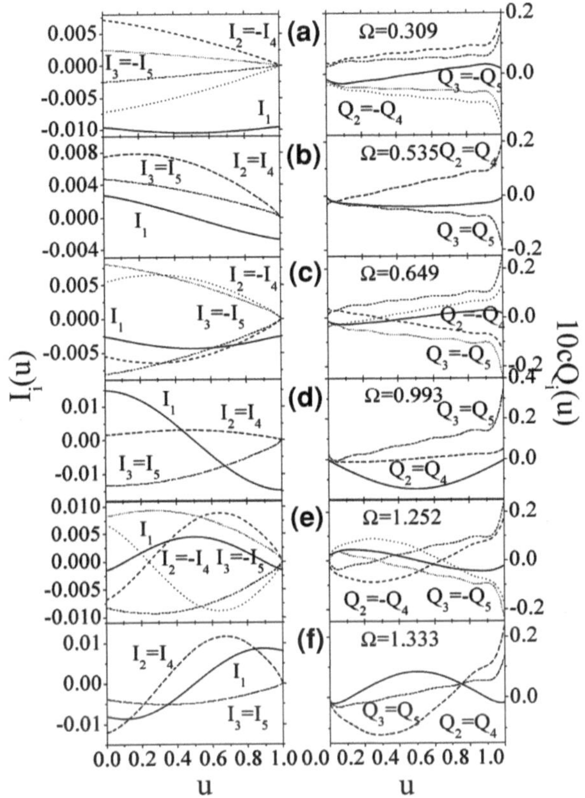

Fig. 5.7 The current and charge distributions for the eigenmodes in an H-shape wire structure with $d_1 = d_2 = 2d_3 = d_4 = 2d_5 = 20$ mm. The other notations are the same as Fig. 5.5 (reprinted with permission from Ref. [4]; copyright (2009), American Institute of Physics)

in Fig. 5.8a–f, $I_2(u) = \pm I_5(u)$, and $I_3(u) = \pm I_4(u)$ are generally preserved by the symmetry requirement. Similar analysis can be done with respect to the charge and current profiles to extract the EM property of the eigenmodes. The outcome of the analysis is summarized below: mode (a), an electric dipole and a magnetic quadrupole; mode (b), an electric quadrupole and a magnetic dipole; mode (c), an electric dipole and a magnetic quadrupole; and mode (d), an electric quadrupole and magnetic dipole. The other modes (e) + (f) are the high frequency counterparts of modes (a) and (d). Our study suggests that the electric and magnetic responses can be visualized most clearly from the charge and current patterns of the eigenmodes, thus the "circuit theory" offers the first hand information which can then be followed by more comprehensive calculations for interesting structures.

From the above analysis, one finds that the eigenfrequency is ultimately determined by the natural boundary conditions at wire ends, thus the wavelengths for these resonances are always compatible to the various lengths of the paths

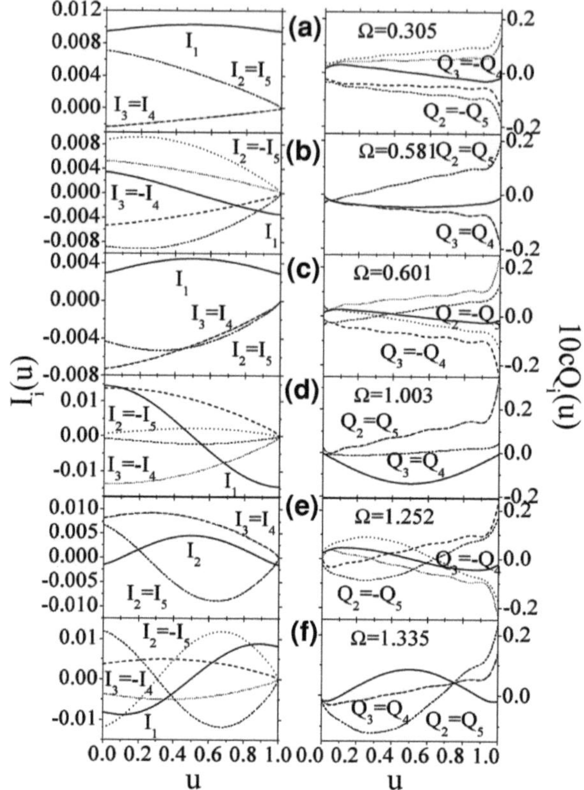

Fig. 5.8 The current and charge distributions for the eigenmodes in an H-shape wire structure with $d_1 = d_2 = 2d_3 = 2d_4 = d_5 = 20$ mm. The other notations are the same as Fig. 5.5 (reprinted with permission from Ref. [4]; copyright (2009), American Institute of Physics)

connected by the wires. To further reduce the eigenfrequency, one easy way of doing this is to consider the doubly stacked T-shape or H-shape circuits and let the separation between the two be as small as possible. It is easy to show that the mutual inductance in such case is logarithmically divergent, the coupling between the two circuits lifts the degeneracy of the two resonances, the "bonding" mode will be pushed down while the "antibonding" mode will be essentially unchanged.

5.4.1 Physical Picture

Again, the physics behind the nature of the symmetry of the solutions is captured in the simplest approximation to the solution of the circuit equation where we ignore the mutual inductance and capacitance and keep only the diagonal terms in the Fourier transform of the circuit parameters. From Eq. (5.20a) with $E^{ext} = 0$, we get

$$I_j(h'_j) = \sum_{k,h_k} X_{jk}^{-1}(h'_j, h_k)(-1)^{s_k}[-(-1)^{h_k}\sigma'_k + \delta_{k=1,2,3}\sigma'_0 + \delta_{k=4,5}\sigma'_1]/\omega.$$

These can be substituted into the boundary conditions (5.21). For example, for those involving the free ends, we get

$$\sum_{h'_j}(-1)^{h'_j}\sum_{k,h_k} X_{jk}^{-1}(h'_j, h_k)[-(-1)^{h_k}\sigma'_k + \delta_{k=1,2,3}\sigma'_0 + \delta_{k=4,5}\sigma'^{s_k}_1] = 0,$$

$$\text{for } j = 2,3,4,5.$$

This can be written as

$$\sum_k Y(j,k)(-1)^{s_k}\sigma'_k = Z_a(j)\sigma'_0 + Z_b(j)\sigma'_1,$$

where

$$Y(j,k) = \sum_{h',h}(-1)^{h'+h}X_{jk}^{-1}(h,h'),$$

$$Z_a(j) = \sum_{k=1,2,3,h',h}(-1)^{h'+s_k}X_{jk}^{-1}(h',h),$$

$$Z_b(j) = \sum_{k=4,5,h',h}(-1)^{h'+s_k}X_{jk}^{-1}(h',h).$$

In our approximation, this becomes

$$y\sigma_j = z\sigma'_0, \quad \text{for } j = 2,3, \tag{5.22a}$$

$$y\sigma_j = z\sigma'_1, \quad \text{for } j = 4,5, \tag{5.22b}$$

where now $z = \sum_h(-1)^h X^{-1}(h,h)$, $y = \sum_h X^{-1}(h,h)$.

Similarly, from the boundary conditions at the two junctions with three wires coming together, we also have

$$3y\sigma'_0 = z[\sigma'_1 + \sigma'_2 + \sigma'_3], \tag{5.23a}$$

$$3y\sigma'_1 = z(\sigma'_5 + \sigma'_4 + \sigma'_0). \tag{5.23b}$$

We have two possibilities:

(1) $y = 0$. In this case we find that $\sigma_0 = \sigma_1 = 0$, $\sigma_2 = -\sigma_3$, $\sigma_4 = -\sigma_5$. This corresponds to the resonances 2, 3.

(2) $y \neq 0$. In this case, as explained below we find that there are two possibilities: (a) $\sigma_{2,3}/\sigma_0 = \sigma_{4,5}/\sigma_0 = \pm 1$. The current at the boundary can be zero. This corresponds to resonance 5. (b) $\sigma_{2,3}/\sigma_0 = \sigma_{4,5}/\sigma_0 = \pm 3/2$. This corresponds to resonance 1 (+ sign) and resonances 4, 6 (− sign). We explain next how we arrive at these conclusions.

Substituting Eq. (5.22a) into Eq. (5.23a), we get

$$(2z^2/y - 3y)\sigma_0' = -z\sigma_1'.$$

y appears in a denominator, this is allowed because $y \neq 0$.

Similarly, from Eq. (5.22b) and (5.23b) we get

$$(2z^2/y - 3y)\sigma_1' = -z\sigma_0'.$$

Combining, we get

$$(2z^2/y - 3y)^2 = z^2.$$

$$2z^2/y - 3y = \pm z,$$

or $2x^2 - 3 = \pm x$ with $x = z/y$.

(a) We first look at the case with the positive sign on the right. Factoring, we get $(x+1)(2x-3) = 0$. For the first root $x = -1$, we get $\sigma_0' = -\sigma_1'$. For $j = 2, 3, 4,$ 5 we get $I_j(x = 0) \propto (-x^2 + 1)\sigma_j'$. Hence for $x = \pm 1$, the currents at the boundaries are zero.

For the second root $x = 1.5$, we get $|\sigma_k'| = 1.5|\sigma_0|$. This condition corresponds to $2[1/X(0) - 2/X(1)] \approx 3[1/X(0) + 2/X(1)]$. Substituting in the expression for X, we get $\pi^2 c^2/[(1 + 10L(0)/L(1))d^2] = \omega^2$. For $L(0) = 86.17$, $L(1) = 73.22$, $L(2) = 57.48$, $\omega^2 = 0.0783\pi^2 c^2/d^2$, $\Omega = 0.28$, $\lambda/d = 7.15$ close to the numerical value in Table 5.2 where we get $\Omega = 0.275$.

(b) For the negative sign, we get upon factoring $(x-1)(2x+3) = 0$. The solution with $x = 1$ corresponds to $\sigma_0' = \sigma_1'$, all the σ_k' for $k = 2, 3, 4, 5$ are the same. $x = -1.5$ corresponds to the dark mode.

5.5 Comparison Between Results in Previous Sections and Results from FDTD Simulations

In this section, we summarize the results by C. Qu and L. Zhou (unpublished data) to compare the theoretical predictions presented above and numerical calculations based on FDTD simulations on the resonance properties of the T and the H metallic wire structures.

5.5.1 Descriptions of the FDTD Simulations

QZ employed FDTD simulations to probe all the resonant modes of metallic wire structures. In their simulation, they set up a special waveguide with a $L \times L$ square cross-section as described in Fig. 5.9, in which two metallic walls are assumed as perfect electric conductors (PEC) while the other two as perfect magnetic

Fig. 5.9 Geometry adopted
in FDTD simulations

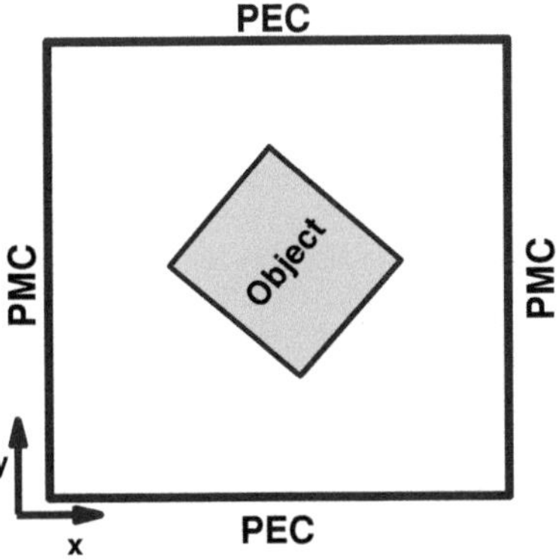

conductors (PMC). The EM boundary conditions are $\hat{n} \times \vec{E} = 0$, $\hat{n} \cdot \vec{B} = 0$ for PEC and $\hat{n} \times \vec{B} = 0$, $\hat{n} \cdot \vec{E} = 0$ for PMC. They then put the metallic wire structure under study into the waveguide (denoted by "object"), and study the transmission spectra through the waveguide under different mode excitations. In principle, a dip appears in the transmission spectrum whenever a resonance mode is excited by the input wave. However, in some cases when a certain resonance mode exhibits good symmetry so that it does not have an overlap with the input electric field, such a mode cannot be excited. In order to probe all resonance modes of a certain structure, QZ rotated the structure by 45° to break the symmetry, and employed multiple waveguide modes (i.e., TEM, TE10, TM10, etc.,) to independently excite the structure. The PEC/PMC walls will inevitably introduce "virtual" images for the objects, so that the couplings between the "objects" and their images may influence the resonance mode frequencies. However, QZ varied the waveguide size and found that such effects are weak. In fact, in the calculations shown below, $L = 80$ mm is typically 3–8 times the wire length, so that the image effect is expected to be weak. Finally, since it is difficult to study a metallic wire with a circular cross-section in FDTD simulations, QZ set the cross-section of each wire as a 0.2×0.2 mm^2 square.

Figures 5.10, 5.11, 5.12 show the FDTD calculated transmission spectra for different T-shaped structures, while Figs. 5.13, 5.14, 5.15, 5.16 show the spectra for different H-shaped structures. The resonance frequencies of these structures obtained by the mode-expansion theory (see previous sections) are shown as the dashed lines in the same figures for direct comparison. The agreement between the two approaches is excellent, which unambiguously validate the developed mode-expansion theory. In those graphs, the unit of length is mm and the unit of frequency is $\omega_u = \pi c / a$ in which a is the length of one of the wires inside the structure.

5.5.2 T-Shape Structures

See Figs. 5.10, 5.11, 5.12.

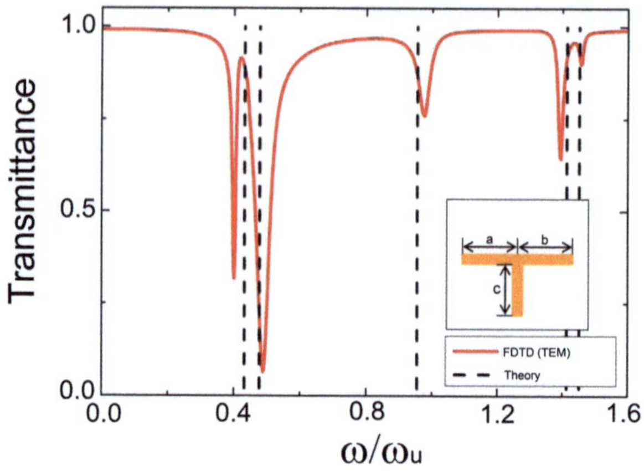

Fig. 5.10 FDTD calculated transmission spectra for EM waves through a waveguide loaded with a T-shaped metallic wire structure with parameters $a = b = c = 10$. Frequencies marked by the *dashed lines* are 0.432, 0.478, 0.955, 1.413, 1.451, which are obtained by the mode-expansion theory

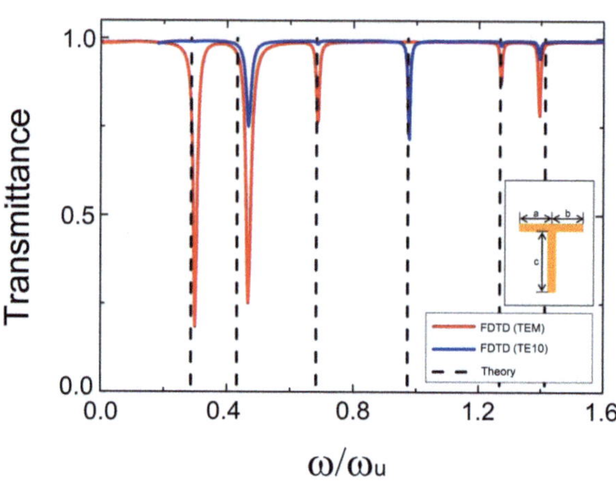

Fig. 5.11 FDTD calculated transmission spectra for EM waves through a waveguide loaded with a T-shaped metallic wire structure with parameters $a = b = 10$, $c = 20$. Frequencies marked by the *dashed lines* are 0.287, 0.432, 0.683, 0.973, 1.268, 1.413, which are obtained by the mode-expansion theory

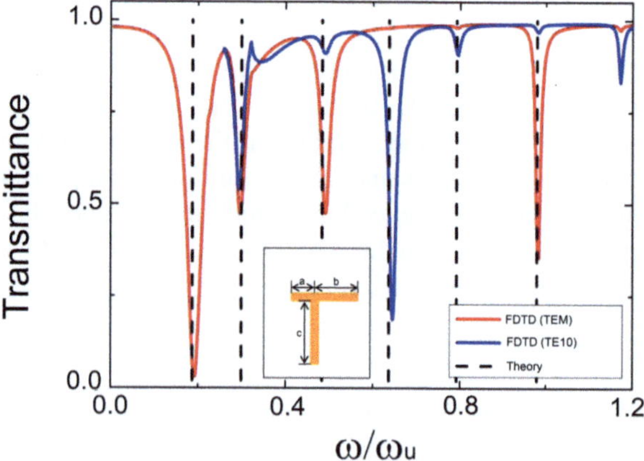

Fig. 5.12 FDTD calculated transmission spectra for EM waves through a waveguide loaded with a T-shaped metallic wire structure with parameters $a = 10$, $b = 20$, $c = 30$. Frequencies marked by the *dashed lines* are 0.187, 0.299, 0.484, 0.637, 0.792, 0.979, which are obtained by the mode-expansion theory

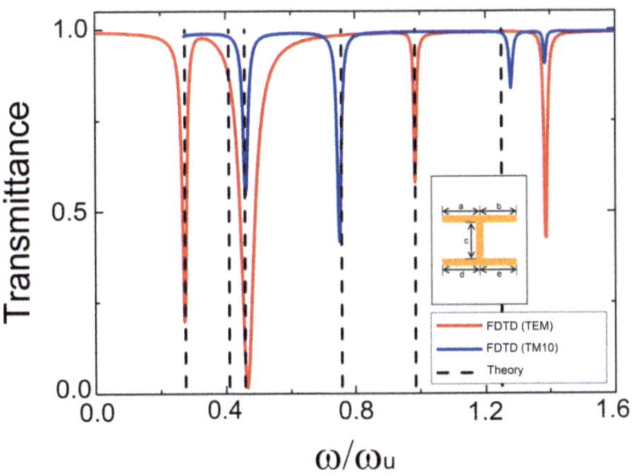

Fig. 5.13 FDTD calculated transmission spectra for EM waves through a waveguide loaded with a H-shaped metallic wire structure with parameters $a = b = c = d = e = 10$. Frequencies marked by the *dashed lines* are 0.275, 0.410, 0.460, 0.758, 0.985, 1.248, which are obtained by the mode-expansion theory

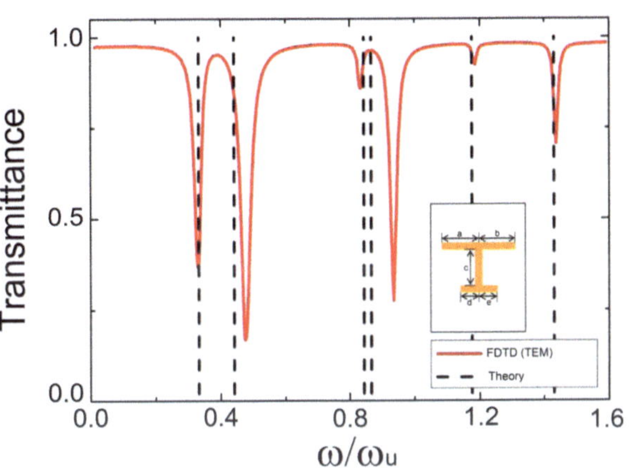

Fig. 5.14 FDTD calculated transmission spectra for EM waves through a waveguide loaded with a H-shaped metallic wire structure with parameters $a = b = c = 20$, $d = e = 10$. Frequencies marked by the *dashed lines* are 0.332, 0.442, 0.844, 0.866, 1.176, 1.431, which are obtained by the mode-expansion theory

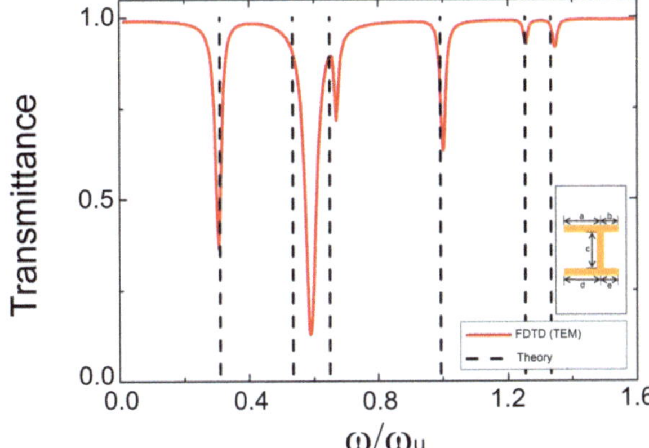

Fig. 5.15 FDTD calculated transmission spectra for EM waves through a waveguide loaded with a H-shaped metallic wire structure with parameters $a = d = c = 20$, $b = e = 10$. Frequencies marked by the *dashed lines* are 0.309, 0.535, 0.649, 0.993, 1.252, 1.333, which are obtained by the mode-expansion theory

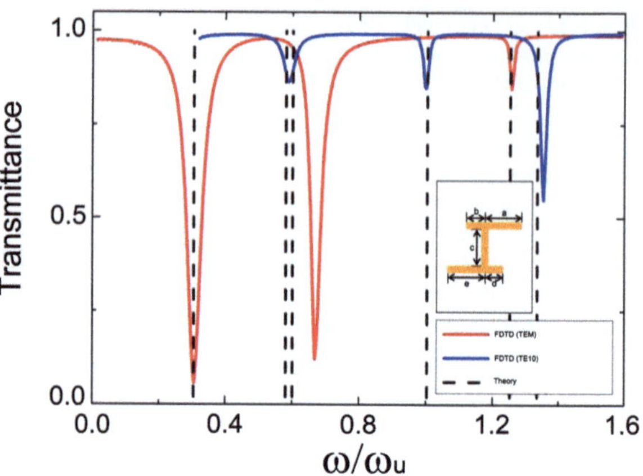

Fig. 5.16 FDTD calculated transmission spectra for EM waves through a waveguide loaded with a H-shaped metallic wire structure with parameters $a = e = c = 20$, $d = b = 10$. Frequencies marked by the *dashed lines* are 0.305, 0.581, 0.601, 1.003, 1.252, 1.335, which are obtained by the mode-expansion theory

5.5.3 H-Shape Structures

See Figs. 5.13, 5.14, 5.15.

5.6 Summary

In this chapter, our equivalent circuit theory has been extended to wire structures with branches. This is achieved with natural boundary conditions and boundary fields at wire ends and joints as the "Lagrange multipliers". The generalized equivalent circuit theory can deal with arbitrary metallic wire networks. Circuit currents, charge profiles as well as boundary electric fields are all self-consistently determined. The application to T-shape and H-shape metallic wire networks suggests that basic types of resonances are mainly determined by symmetry and wire dimensions of a network. The results are well represented by a simple approximation ignoring the mutual circuit parameters and including only the diagonal Fourier components. The low frequency modes take place along the longer wire paths while the higher frequency modes can be generated via adding nodes in the circuit patterns along various wire paths. The characteristic behavior of the electric and magnetic responses can be visualized from the charge and circuit current profiles for a given mode.

We have discussed a way to represent our physical variables in a Fourier series. In this way the extended function is discontinuous at the boundaries. For a function $f(x)$ that is discontinuous at isolated points, the Fourier series converges to $[f(x+) + f(x-)]/2$ at these points. Thus, the Fourier series for the currents I only converges to a value $I/2$ at the boundaries. The only places where these boundary values are important are in the current conservation boundary conditions. Fortunately the current conservation condition is homogeneous. The condition $\sum_j I_j = 0$ is the same as the condition $\sum_j I_j/2 = 0$. Thus the correct equations are recovered.

Finally, we compared our theoretical approach with full-wave FDTD simulations on certain T- and H-shaped metallic wire structures, and found excellent agreement between these two approaches. Such a comparison fully justified the theoretical approach developed in this chapter.

References

1. Z. Liu, X. Zhang, Y. Mao, Y.Y. Zhu, Z. Yang, C.T. Chan, P. Sheng, Science **289**, 1734 (2000)
2. W. Wen, L. Zhou, J. Li, W. Ge, C.T. Chan, P. Sheng, Phys. Rev. Lett. **89**, 223901 (2002)
3. Y. Kosako, F. Kadoya, H.F. Hofmann, Nat. Photon. **4**, 312 (2010)
4. W. Zhang, S.T. Chui, J. Appl. Phys. **105**, 113121 (2009)

Chapter 6
Jerusalem Cross

6.1 Introduction

The behavior of electrical currents in wire networks in the zero frequency limit is governed by Kirchoff's law. We have discussed, in previous chapters, how to generalize this to the finite frequency case where the currents are no longer uniform along the wires. We found that it is necessary to introduce electric fields σ_i localized at the corners and the junctions of the network. As the frequency is increased, the network exhibits a series of resonances. We found that the patterns of the currents of the resonances are completely characterized by the electric fields at the junctions. In this chapter, we consider more complicated structures such as the Jerusalem Cross (JC) [1]. We hope to provide enough detail for this case that the reader will be able to learn and deal with any general wire structures.

Together, with results discussed in previous chapters, our calculation suggests that for symmetrical structures there are in general "sets" of resonance patterns. The resonance patterns in each "set" consist of higher harmonics of a fundamental pattern. Among the different "sets" there are some in which groups of the local electric fields become zero. The number of such groups is given by the number of groups of inequivalent vertices. We further illustrate this with the JC structure in this chapter.

The JC structure is shown in Fig. 6.1. To be specific, we consider an example so that each section of the wire is of length d and width a with $d/a = 100$.

There are 12 wires/currents, 13 vertices. There are three groups of inequivalent vertices:

(a) The end vertices 5, 6, 7, 8, 9, 10, 11, 12.
(b) The "branch" vertices 1, 2, 3, 4 where three wires come together;
(c) The middle vertex 0.

Any function $F(x)$ can be expanded as $F(x) = \sum_h F(h) \exp\left(+\frac{i\pi hx}{d_k}\right)$, where h is an integer and $x_k - x_{o(k)} = d_k$ is the wire length. Here $o(k)$ denotes the other end of

S. T. Chui and L. Zhou, *Electromagnetic Behaviour of Metallic Wire Structures,*
DOI: 10.1007/978-1-4471-4159-4_6, © Springer-Verlag London 2013

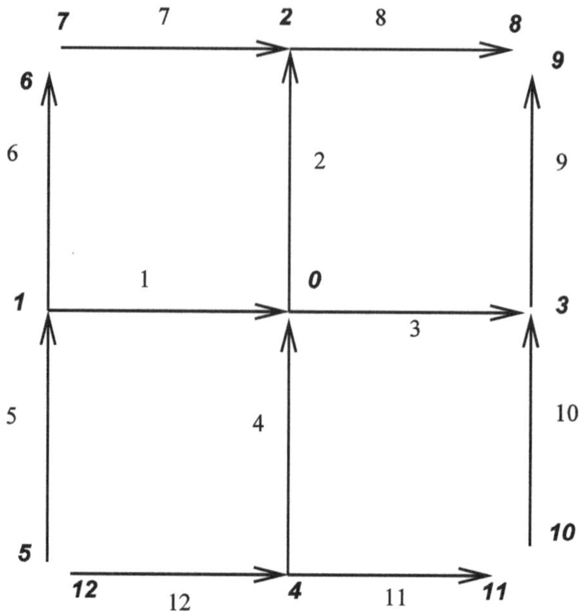

Fig. 6.1 The Jerusalem Cross with labels for the wires and the vertices. (Reprinted with permission from Ref. [1]. Copyright (2009), American Institute of Physics.)

vertex k. The value of the function at the ends is given in terms of the Fourier transform as $F(x = 0) = \sum_h F(h); \ F(x = d) = \sum_h (-1)^h F(h)$.

The currents I_j are related to the external electric field E_{ext} and the local electric field σ at the corners by circuit equations of the form

$$E'_{ext}(h_k) = \sum_{j,h'_j} X_{kj}(h_k, h'_j) I_j(h'_j) + (-1)^{s_k} [-\frac{(-1)^{h_k}}{\omega} \sigma'_k + \frac{\sigma'_x}{\omega}]. \qquad (6.1)$$

where X is the impedance matrix. Here x, k denote the beginning and the end point where we perform the Fourier transform. Examples of this for the T and the H structures were given in Eqs. (5.16) and (5.20). In our example, we shall set $(x, k) = (1,5); (1,6); (2,7); (2,8); (3,9); (3,10); (4,11); (4,12)$ for the wires on the perimeter and $(x, k) = (0,1); (0,2); (0,3); (0,4)$ for the wires connecting the middle. In our example, the currents go from x to k. We have set a sign convention so that a current going up is positive. The sign $(-1)^{s_k}$ in Eq. (6.1) takes care of this convention. For the above two sets of vertex pairs we have $s_k = 0, 1, 0, 1, 1, 0, 1, 0$ for the perimeter wires and $s_k = 0, 1, 1, 0$ for the middle wires. This is summarized in Table 6.1.

For example, for the vertex pair $(1, 5)$ corresponding to the 5th column in the above table and the lower left wire segment in Fig. (6.1), the segment index corresponds to 5. We get the equation

Table 6.1 Perimeter and middle wires for two sets of vertex pairs

k	1	2	3	4	5	6	7	8	9	10	11	12
s_k	0	1	1	0	0	1	0	1	1	0	1	0
x	0	0	0	0	1	1	2	2	3	3	4	4

$$E'_{ext,5}(h_5) = \sum_{j,h'_j} X_{5j}(h_5,\, h'_j)I_j(h'_j) + [-\frac{(-1)^{h_5}}{\omega}\sigma'_5 + \frac{\sigma'_1}{\omega}].$$

For a resonance, $E_{ext} = 0$ the resonance frequencies are determined by the equation

$$I(h') = \sum_{j,h'_j} X_{jk}^{-1}(h',\, h)(-1)^{s_k}[-\frac{(-1)^{h_k}}{\omega}\sigma'_k + \frac{\sigma'_x}{\omega}]. \tag{6.2}$$

There are three classes of current conservation boundary conditions associated with the three classes of inequivalent vertices:

(a) The currents at the free ends of wires $i = 5, 6, 7, 8, 9, 10, 11, 12$ (vertex type A) are zero.

$$\sum_{h_i}(-1)^{h_i}Ii(h_i) = 0,\,; \tag{6.3a}$$

(b) The currents at the "branch" vertices of type B are conserved:

$$\sum_{h}[(-1)^h I_1(h) - I_5(h) + I_6(h)] = 0. \tag{6.3b}$$

$$\sum_{h}[(-1)^h I_2(h) + I_7(h) - I_8(h)] = 0.$$

$$\sum_{h}[(-1)^h I_3(h) + I_{10}(h) - I_9(h)] = 0.$$

$$\sum_{h}[(-1)^h I_4(h) + I_{11}(h) - I_{12}(h)] = 0.$$

(c) The currents going into the middle vertex (C) are conserved:

$$\sum_{h}[I_1(h) + I_4(h) - I_2(h) - I_3(h)] = 0. \tag{6.3c}$$

Substituting Eq. (6.2) into Eq. (6.3a–6.3c), we obtain a set of equations of a form $H_0 Y^{-1}\sigma = 0$ where the matrix $H_0 Y^{-1}$ is a nonlinear function of the frequency ω. These frequencies can be solved by the method discussed in the previous chapter: We solve the eigenvalue equation $H_0 Y^{-1}\sigma = \lambda\sigma$ for different values of ω. A resonance is indicated when the number of eigenvalues of positive

Table 6.2 The local electric fields for the Jerusalem Cross at different resonant frequencies ω in units of c/d_1

	$\omega = 0.113$	$\omega = 0.12$	$\omega = 0.133$	$\omega = 0.373$	$\omega = 0.83$	$\omega = 0.875$	$\omega = 0.88$
σ_1	(0.24 0.11)	(0.15 −0.07)	(−0.06 0.28)	(−0.13 0.00)	(0.00 0.01)	(0.00 0.00)	(−0.01 0.00)
σ_2	(−0.07 −0.28)	(0.01 0.37)	(0.08 −0.29)	(−0.14 0.00)	(0.00 0.01)	(0.00 0.00)	(0.01 0.00)
σ_3	(−0.25 −0.12)	(−0.16 0.06)	(−0.08 0.27)	(−0.14 −0.01)	(0.00 −0.01)	(0.00 0.00)	(0.01 0.00)
σ_4	(0.07 0.29)	(0.00 −0.36)	(0.08 −0.26)	(−0.14 0.01)	(0.00 −0.01)	(0.00 0.00)	(−0.01 0.00)
σ_5	(0.25 0.13)	(0.15 −0.10)	(−0.07 0.29)	(−0.18 0.00)	(−0.11 0.36)	(0.29 −0.26)	(−0.29 −0.24)
σ_6	(0.24 0.08)	(0.15 −0.04)	(−0.07 0.29)	(−0.17 0.00)	(0.10 −0.32)	(−0.25 0.22)	(0.25 0.20)
σ_7	(−0.05 −0.28)	(0.02 0.37)	(0.08 −0.30)	(−0.18 0.00)	(0.09 −0.32)	(0.24 −0.22)	(−0.26 −0.20)
σ_8	(−0.09 −0.30)	(0.00 0.38)	(0.08 −0.30)	(−0.19 0.00)	(−0.11 0.36)	(−0.29 0.26)	(0.30 0.24)
σ_9	(−0.26 −0.15)	(−0.17 0.10)	(−0.08 0.27)	(−0.18 −0.01)	(0.11 −0.36)	(0.29 −0.24)	(0.30 0.24)
σ_{10}	(−0.25 −0.10)	(−0.17 0.03)	(−0.08 0.27)	(−0.17 −0.01)	(−0.10 0.32)	(−0.24 0.20)	(−0.26 −0.21)
σ_{11}	(0.05 0.29)	(−0.01 −0.37)	(0.08 −0.26)	(−0.17 0.01)	(−0.09 0.32)	(0.24 −0.20)	(0.25 0.21)
σ_{12}	(0.10 0.31)	(0.01 −0.37)	(0.08 −0.26)	(−0.18 0.01)	(0.11 −0.36)	(−0.28 0.24)	(−0.29 −0.24)
σ_0	(0.00 0.00)	(0.00 0.00)	(0.00 0.00)	(0.82 0.00)	(0.00 0.00)	(0.00 0.00)	(0.00 0.00)

real parts is changed. Carrying out this calculation, we arrive at our result summarized in Table 6.2 where we show the boundary electric fields and the resonance frequencies for the low lying resonance modes. (The overall phase of the eigenfunction σ is arbitrary.)

The current directions of the first 6th resonances quoted in the above table are illustrated in Figs. 6.2 and 6.3.

As can be seen from Table 6.2, among the different resonance modes there are some, in which, groups of the local electric fields become zero within the numerical accuracy of our calculation. There are three classes of solutions:

1. Only the local field at vertex 0 (type A) is zero ($\omega = 0.113, 0.12$),
2. the local fields at vertices 0, 1, 2, 3, 4 (type B and C) are zero ($\omega = 0.83, 0.875, 0.88$).
3. all local fields at the internal junctions are nonzero ($\omega = 0.373$).

The local electric fields at the free ends are never zero, however!

This kind of behavior is captured in the simplest approximation to the solution of the circuit equation where we ignore the mutual inductance and capacitance and keep only the diagonal terms in the Fourier transform of the circuit parameters.

Fig. 6.2 The configuration of the first three normal modes

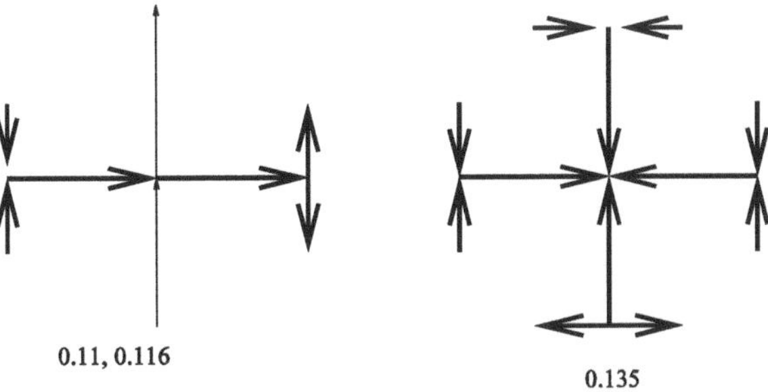

Fig. 6.3 The configuration of the 3rd to 6th normal modes

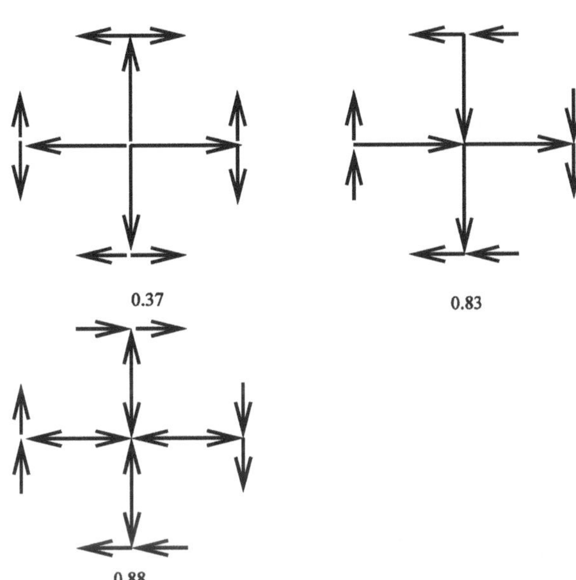

Examples of these have been given in Chap. 5. To further reinforce the basic physical ideas, we describe this next.

In this simple approximation, the current for each Fourier mode in a wire segment is determined by the sum of the electric fields at the two ends divided by the impedance; we get

$$I_j(h) = X^{-1}(h, h)(-1)^{s_j}[(-1)^h \sigma'_j - \sigma'_{x(j)}]/\omega. \tag{6.4}$$

Here, the symbol $x(j)$ indicates the origin of the corresponding jth segment. The boundary condition of current conservation at the type (A) vertices, Eq. (6.3a), become

$$\sum_h I_j(h)(-1)^h = 0.$$

With $z = \sum_h (-1)^h X^{-1}(h, h)$, $y = \sum_h X^{-1}(h, h)$, this current conservation condition at the class (A) vertices j can be written as

$$y\sigma_j' = z\sigma_{x(j)}'. \tag{6.4a}$$

Similarly, the current conservation condition at the class (B) vertices, Eq. (6.3b), becomes

$$-3y\sigma_1' + z(\sigma_0' + \sigma_5' + \sigma_6') = 0. \tag{6.4b}$$

with similar equations involving $\sigma_{2,3,4}'$. Finally, for the class (C) vertex boundary condition, Eq. (6.3c), we get

$$-z\sigma_1' + 4y\sigma_0' - z\sigma_2' - z\sigma_3' - z\sigma_4' = 0. \tag{6.4c}$$

From equation (6.4a) we get $y\sigma_{5,6}' = z\sigma_1'$, etc. Hence, either $y = 0$, $\sigma_{1,2,3,4}' = 0$. or $y \neq 0$, $\sigma_{5,6}' = z\sigma_1'/y$. etc. For the first case from Eq. (6.4b) we find that $\sigma_0 = 0$. This provides for a simple way to understand the type (1) solutions that we see from our numerical calculation. For the second possibility, from Eq. (6.4b) we get $(3y - 2z^2/y)\sigma_1' = z\sigma_0'$. Hence, either $3y - 2z^2/y = 0$, $\sigma_0' = 0$; or $3y - 2z^2/y \neq 0$, $\sigma_1' = z\sigma_0'^2/y$. These two types of solutions correspond to the class (2) and class (3) solutions that are shown in Table 6.2.

To summarize, we get

1. $3y - 2z^2/y = 0$, $\sigma_0 = 0$; or
2. $y = 0$, $\sigma_{1,2,3,4} = 0$, $\sigma_0 = 0$; or
3. $\sigma_1' = z\sigma_0'^2/y$.

This illustrates the physical origin of the resonance modes, where some of the boundary electric fields at the junctions become zero.

Similar results have already been discussed for the T, H structures in the previous chapter. We have also looked at the simpler "cross structure" which consists of vertices: (1) 1, 2, 3, 4 (2) 0 in Fig. 6.1. The corresponding class of resonance modes is the following: (1) all fields are nonzero, (2) the field at vertex 0 is zero.

Reference

1. S.T. Chui, W.Y. Zhang, J. Appl. Phys. **106**, 074904 (2009)

Chapter 7
Metallic Wire Structures Under a Moderate Electromagnetic Field

Metallic wire structures from simple [1] to complex [2–4] have long been used as absorbers of electromagnetic radiation. As we discussed in previous chapters, the response of the wire structure is characterized by a series of resonances. In our approach, we introduce new variables which consist of localized electric fields at the ends and junctions of the wire structures. Near the resonance frequencies for optimum absorption the electric fields at the ends of the wire structure are generally larger by many orders of magnitude than the magnitude of the external electric field. This absorption application is limited by the magnitude of the current the wire can carry before dielectric breakdown at the ends of the wire occurs [5]. The maximum current is surprisingly small. The dielectric breakdown field in air σ_b, is 3×10^6 V/m, a field of the order of 1 V/m applied to the body of the wire cause such a large field at the ends and induce a breakdown. The approach we use in the description of the wire structure is particularly well suited to the study of this problem. In this chapter, we shall clarify why the breakdown current is so small and describe how this can be computed. Some suggestions for improved absorbers are discussed. For example, the maximum absorption of the double split ring is higher than that of a single ring by an order of magnitude.

To illustrate the crucial physical ideas, we first consider a simple structure, the single split ring. The results for a straight wire are essentially the same. Take the coordinate system so that the z axis is perpendicular to the ring. Assume a time dependence of the form $\exp(i\omega t)$. We expand the current along the ring in a Fourier series as $I(\phi) = \sum_m I_m \exp(im\phi)$. The total impedance for the mth Fourier mode, X_m, is given by Eq. (2.17):

$$X_m(\omega) = r_c + iY_m; \tag{7.1}$$

$Y_m = L_m \omega - m^2/(\omega \bar{C}_m)$; L_m, $m^2/\bar{C}_m$ are the Fourier components of the self-inductances and capacitances. r_c is the resistance per unit length of the conductor. r_c is usually less than 1 Ω, much less than Y_m, which is of the order of the permittivity of the vacuum, $Z_0 = (\mu_0/\epsilon_0)^{1/2} = 377\,\Omega$, divided by the linear size of the system. At higher frequencies, there will be additional contribution from the

S. T. Chui and L. Zhou, *Electromagnetic Behaviour of Metallic Wire Structures*,
DOI: 10.1007/978-1-4471-4159-4_7, © Springer-Verlag London 2013

radiation resistance, as is explained in the Appendix of Chap. 2. One might have thought that the maximum magnitude of E_{ext} is the dielectric breakdown field σ_b and the maximum power that can be absorbed is of the order of $\sigma_b^2 r_c$. As we explained below, near a resonance, the ratio of the boundary electric field to the external field σ/E_{ext} is of the order of Y/r_c. A dielectric breakdown occurs when the boundary electric field is of the order of σ_b and the external field of the order of $\sigma_b r_c/Y$. The actual maximum power that can be absorbed, E_{ext}^2/r_c, is much lower, of the order of $\sigma_b^2 r_c/Y^2 \sim \sigma_b^2 r_c/Z_0^2$. Thus, one way to absorb more power is to lower the effective impedance, so that Z_0 is replaced by a smaller number, as in a double split ring. We now explain this in detail.

In our treatment of the split ring, we introduce the localized electric field at the ends of the ring that is given by Eq. (3.6):

$$\sigma = -\sum_m E_{\text{ext}}^m/X_m / \sum_m 1/X_m. \tag{7.2}$$

The resonance frequency ω_c is determined from the condition that a nontrivial solution still exist even when $E_{\text{ext}} = 0$. In previous chapters, we have not been careful about the complex nature of the impedance. When the resistance is included, the resonance condition requires only that the imaginary part of the impedance becomes zero. Since $r_c \ll Y_m$, $\text{Im}[1/X_m(\omega)] \sim -1/Y_m$, the resonance condition becomes:

$$\sum_m \text{Im}[1/X_m(\omega_c)] = -\sum_m 1/Y_m(\omega_c) = 0.$$

The real part of the inverse impedance, $\text{Re}[1/X_m(\omega)] \sim r_c/Y_m^2$, is much smaller. At resonance, we get

$$\sum_m 1/X_m(\omega_c) = \sum_m r_c/Y_m^2(\omega_c) - i\sum_m 1/Y_m(\omega_c) \sim \sum_m r_c/Y_m^2(\omega_c). \tag{7.3}$$

We next determine the current induced in the wire structure at resonance.

Generally, the external electric field is slowly varying compared with the size of the system. To illustrate the essential physics, we shall only include the slowest varying component of the electric field. A vector electric field induces only the $m = \pm 1$ components, whereas a spatially varying magnetic field can induce an $m = 0$ component of the external field. We illustrate our result by keeping only the $m = 0$ component of the external field electric field. Similar result is obtained for the $m = \pm 1$ case. In this approximation from Eq. (7.2) we obtain

$$E_{\text{ext}} = \alpha\sigma \tag{7.4}$$

with

$$\alpha = -X_0 \sum_m 1/X_m \tag{7.5}$$

since $\sum_m E^m_{\text{ext}}/X_m \sim E_{\text{ext}}/X_0$. The maximum breakdown external field that can be applied to the wire before the localized field at the ends reach the breakdown value $\sigma_b = 3 \times 10^6$ V/m is given by

$$E_{\text{ext},b} = \alpha\sigma_b$$

For coupling to the $m = \pm 1$ components of the electric field, we would have $\alpha = -X_{\pm 1}\sum_m 1/X_m$ instead of Eq. (7.5). Near a resonance, α is very small. The fields at the ends are much larger than the external field. Although σ_b is large, the actual external electric field that can be applied is smaller by several orders of magnitude. More precisely, at a resonance, from Eq. (7.3) and (7.5) we get

$$\alpha = -X_0 \sum_m r_c/Y^2_m. \tag{7.6}$$

Since X_0/Y_{m^2} is of the order of $1/(\omega_c L)$, α is of the order of $r_c/(\omega_c L)$ with ω_c of the order of $\omega_0 = c/R$. In the thin-wire limit $L = -\ln(a/R)/(2\pi c^2)$. We obtain $\alpha \approx O(2\pi R r_c/Z_0 \ln(a/R))$. is thus of the order of the resistivity of the ring, $2\pi R r_c$ divided by the permittivity of the vacuum, $Z_0 = (\mu_0/\epsilon_0)^{1/2} = 377\,\Omega$. The resistivity of Cu is of the order of $10^{-6}\,\Omega$cm. For cm size rings made with Cu, $\alpha(\omega_c)$ is thus of the order of 10^{-7}. In this limit, the Fourier component of the maximum critical current in the wire is given by

$$I^c_m = \sigma_b/X_m \quad \text{for } m \neq 0$$

$$I^c_0 = (E^{\text{ext}}_0 + \sigma_b)/X_0.$$

We next examine the consequence of these results.

We first examine the issue of absorption. The ohmic loss, $r_c \sum_m I^2_m$, is equal to $P_b = r_c\left(\sum_{m>0} I^2_m + I^2_0\right) = r_c\sigma^2_b\left(\sum_{m>0} 1/X^2_m + (1+\alpha)^2/X^2_0\right)$. This can be written as

$$P_b = r_c\sigma^2_b\left[\sum_m 1/X^2_m + \alpha(\alpha+2)/X^2_0\right]. \tag{7.7}$$

This is the maximum power that can be absorbed. As is advertised at the beginning of this chapter, the power that can be absorbed is of the order of $\sigma^2_b r_c/Z^2_0$.

For a straight wire of length L, we obtain a similar result where now we expand the physical quantities as a Fourier series along the wire; m is replaced by the wave vector $k_n = 2\pi n/L$.

The antenna can also be used to transmit power. This power can be computed by a multipole expansion. The lowest contribution comes from the dipolar radiation. The power dissipated is proportional to the dipole moment square, which, in turn, is proportional to $I^2_{m=1}$. Thus the dipole antenna cannot transmit very high power.

We next show that the power absorption for a double split ring can be much bigger than that for a single ring because the effective impedance is lowered. We focus on rings such that the positions of the cuts of the two rings are at opposite ends of the diameter. We are interested in the limit when, t, the wire to wire separation between the rings, is small. This system is discussed in Chap. 3. It has been much studied recently in the context of negative refracting material. We argue here that they are useful for high field absorption as well. We call the two rings 1 and 2. As is discussed in Eq. (3.1) and (3.2) the electric field in ring 1 is a sum of a "self" and a "mutual" contribution:

$$E_{1m} = E_{\text{ext},1m} + \sigma_1 = X_m I_{1m} + X'_m I_{2m}$$

where $X_m = r_c + i[L_m \omega - m^2/(C_m \omega)]$, $X'_m = i[L'_m \omega - m^2/(C'_m \omega)]$. L_m, C_m (L'_m) are the self-(mutual) inductances and capacitances. Similarly for ring 2, we get $E_{2m} = E_{\text{ext},2m} + (-1)^m \sigma_2 = X_m I_{2m} + X'_m I_{1m}$. There are optic and acoustic resonance modes, corresponding to

$$\sigma_{1\pm} = \pm \sigma_{2\pm},\ I_{1m} = \pm(-1)^m I_{2m}.$$

Going through the same calculation as the one ring case, we find the resonance condition [Eq. (3.4)]

$$\sum_m 1/[X_m(\omega_c) \pm (-1)^m X'_m(\omega_c)] = 0. \tag{7.8}$$

Now the ohmic loss is given by

$$P_{2b} = r_c \sum_m m(I_{1m}^2 + I_{2m}^2). \tag{7.9}$$

Expressions for the currents are given in Eq. (3.4b) as

$$\begin{aligned} I_{1m} &= [X_m(E_{\text{ext},1m} + \sigma_1) - (E_{\text{ext},2m} + (-1)^m \sigma_2)X'_m]/(X_m^2 - X_m'^2); \\ I_{2m} &= [X'_m(E_{\text{ext},1m} + \sigma_1) - (E_{\text{ext},2m} + (-1)^m \sigma_2)X_m]/(X_m^2 - X_m'^2). \end{aligned} \tag{7.10}$$

We make the same approximation as the single ring case by keeping the slowly varying component of the external field. For magnetic excitations with a dominant $m = 0$ component, the breakdown external field is now given by (see Eq. 3.4d)

$$E_{\text{ext},b}^{\pm} = \alpha_2 \sigma_b, \tag{7.11}$$

where $E_{\text{ext},b}^{\pm} = (E_{\text{ext},1b} \pm E_{\text{ext},2b})/2$,

$$\alpha_2 = -(X_0 \pm X'_0) \sum_m 1/[X_m(\omega_c) \pm (-1)^m X'_m(\omega_c)]. \tag{7.11a}$$

Again, close to the resonance $\alpha_2(\omega_c)$ is small, $E_{\text{ext},b}$ is much smaller than σ_b. When the rings are close together we expect E_1 to be close to E_2 and the plus (minus) sign in the above equation to be applicable for $m = 0$ $(m = \pm 1)$. From Eqs. (7.9–7.10), the ohmic power absorbed is now given by

$$P_{2b} = 2r_c\sigma_b^2\left[\sum_m 1/(X_m + (-1)^m X_m')^2 + \alpha_2^2/(X_0 + X_0')^2\right]. \qquad (7.12)$$

The first term on the right hand side looks similar to that in the resonance condition, Eq. (7.8), but there is a major difference in that the summands are all positive. For m odd, because X_m is close to X_m', the denominator of some of the summands contributing to the first term on the right hand side is close to zero. The quantity $1/(X_{1+2n} - X_{1+2n}')^2$ becomes very big. In the expression for the resonance condition, Eq. (7.8), the expression $1/(X_{1+2n} - X_{1+2n}')$ occurs but only to a first power, there is cancelation among different summands. In the above expression for the power, this cancelation does not occur. Thus, there is a big increase in the power absorbed because the effective impedance is lowered for some of the Fourier modes. For example, in the thin-wire limit the inductances and the capacitances can be analytically evaluated; we find (see the discussion near Eq. (3.4) in Chap. 3) $X_m - X_m' = r_c - i\ln[a/(a+t)](\omega/\omega_0)Z_0/R(1 - m^2\omega_0^2/\omega^2)$, $\omega_0 = c/R$. When the wire resistivity r_c and the separation t are small, P_{2c}/P_c is of the order of $\ln^2[a/(2a+t)]/\ln^2(a/R)$ and can be easily increased by a factor of 10. For example, for $a = t = 0.5$ mm, $R = 11$ mm, this ratio is equal to 7.9.

For electric excitations with a dominant $m = 1$ component, Eq. (7.11) remains valid except that now

$$\alpha_2 = -(X_1 \pm X_1')\sum_m 1/[X_m(\omega_c) \pm (-1)^m X_m'(\omega_c)],$$

$$E_{\pm,1} = (E_{1,1} - \pm E_{2,1})/2.$$

Now the dominant absorption is contributed by the "−" mode.

We finish this chapter by considering the power absorption in the Jerusalem Cross studied in the previous chapter. This is a more complicated multiply connected wire structure where in addition to the free ends there are junctions with at least three wires joining together. We wish to illustrate how, for a general structure, the idea we just discussed for a simple structure remains true.

For the multiply connected wire structures there are local fields not only at the free ends, but at the junctions as well. As we discussed in previous chapters, among the different resonance modes there are some in which the currents along some segments are zero and groups of the local electric fields at the *junctions* at the ends of these segments become zero. The local electric fields at the free ends are never zero; however, the basic idea of a large field at the ends is still applicable.

More precisely, the ohmic loss is the sum of the loss at each of the wire segments: $p = \sum_j r_{cj}I_j^2$. I_j can be written as linear combinations of the external field and the boundary electric fields σ_j, as is illustrated above. While some of the boundary fields can be zero; because, the field at the free boundary is always nonzero, the absorption will be limited by the dielectric breakdown caused by these fields. For this structure, unlike the coupled split rings, there is no *near* cancelation of effective impedance. For example, in the simple approximation described in the previous chapter, the effective impedances are [Eq. (6.4a)] y and z, neither of which is small. We thus expect that, up to a numerical factor of the order of unity, the absorption is the same as that of a single ring/line element.

We have also looked at other structures such as the "T," "H" and the "+" structures in previous chapters. For these cases there are two groups of vertices and we find two classes of solutions where the local fields at the central group of vertices can be finite or zero, as is expected. In all cases, the fields at the free ends are always nonzero. Thus our consideration in this chapter is still applicable.

In conclusion, we study, in this chapter, the limitation of high power absorption of metallic wire structures due to local electric fields at the wire vertices. The electric field that can be applied to the wire structure is much smaller than the dielectric breakdown field in air. No matter what the frequency or shape the wire structure is, the local electric field at the free wire ends are never zero. Nevertheless, the absorption efficiency of wire structures can be improved dramatically, as is demonstrated by comparing the single and double split ring structures.

References

1. J.H. Van Vleck, F. Block, M. Hammermesh, J. Appl. Phys. **18**, 274 (1947)
2. W. Wen, L. Zhou, J. Li, W. Ge, C.T. Chan, P. Sheng, Phys. Rev. Lett. **89**, 223901 (2002)
3. N.I. Landy et al., Phys. Rev. **100**, 207402 (2008)
4. B. Hou, H. Xie, W. Wen, P. Sheng, Phys. Rev. B **77**, 125113 (2008)
5. S.T. Chui, W.Y. Zhang, J. Appl. Phys. **106**, 074904 (2009)

Chapter 8
Electromagnetic Waves in Wire Composites I: Plasmonics

8.1 Introduction

In this chapter and the next we consider the situation where there is a collection of metallic wire structures and examine the propagation of EM waves in them. We consider two regimes, depending on whether the wavelength *outside the metallic structure* is larger or smaller than the average spacing between the structures. (1) In the long wavelength limit, the effect of the structures can be described by effective susceptibilities. (2) The small wavelength limit is more complex and has to be treated numerically. In this book we describe one way to handle this class of problems with the multiple scattering method. At frequencies close to resonances of the basic units, expansion of the fields in *all* regions in terms of plane waves is not very efficient. The multiple scattering approach is better.

At long wavelengths outside the wire structure, the physics is often determined by the effective susceptibilities $\langle \epsilon \rangle$ and $\langle \mu \rangle$ of the basic unit and hence of the composite. One of the issues we shall address is how to determine these susceptibilities. So far we have focused on the variations of the currents along the metallic wires. For some applications, the resonance can come from the variation of the current perpendicular to the wire. For this latter case, we find a "pseudopotential" idea to be useful in determining the effective susceptibilities. The effective susceptibility of the wire structure can then be used to determine that of the composite with existing effective medium theories such as the coherent potential approximation so that the average t matrix (explain below) is zero in the effective medium.

As far as the EM field outside the structure is concerned, the effect of the structure is completely determined by the t matrix which can be expressed in terms of the scattering phase shifts η_n for various eigenfunction index (such as the angular momentum components) n. The behavior of the EM wave outside the structure can be understood *entirely* in terms of the scattering phase shift of the structures. In the pseudopotential idea in electronic structure calculation a real potential can be replaced by an effective one so that the same scattering of the

S. T. Chui and L. Zhou, *Electromagnetic Behaviour of Metallic Wire Structures*,
DOI: 10.1007/978-1-4471-4159-4_8, © Springer-Verlag London 2013

electrons is produced. Similarly effective susceptibilities can be introduced so that the correct scattering effect for electromagnetic waves is produced. In this chapter we give examples of how this can be done.

A class of problems of common interest occurs at frequencies close to the scattering resonances of constituents of these structures. Since the observation of surface enhanced Raman scattering and the transmission of EM waves through subwavelength holes in a metallic film by Ebessen and coworkers [1], there have been much interests in the role play by surface plasmons [2]. Much of the current foci in this area consist of the design of different structures that improve the performance for (a) enhanced interaction of light with surfaces and adsorbed molecules and (b) low-loss subwavelength transport of light. The former can be viewed as improving the scattering cross-section of the relevant combined system (the surface of interest or the hybridized state of the adsorbed molecule and the surface); the latter as enabling the dispersion at a frequency of interest so that the wave vector perpendicular to (along) the direction of propagation is imaginary (as real as possible). In this chapter, we hope to provide some simple fundamental physics and examples that are connected with these current foci of application. Because the "strong coupling" limit is the focus of interest, perturbation theory usually is a poor approximation for this class of problems. The surface plasmon resonance also corresponds to a scattering resonance. According to the pseudo-potential idea discussed above, the same physics is produced if the metal is replaced by other material as long as the scattering phase shift is the same. Thus, plasmonics is not necessarily restricted to metals but can be considered the study of effect of scattering resonances.

With the pseudopotential idea we obtain a dielectric constant for a metallic cylinder of the form of a simple metal

$$\varepsilon_{\text{eff}} = 1 - \omega_p'^2 / \omega^2 \tag{8.1}$$

where the effective "plasma frequency" is given by

$$\omega_p'^2 = -\frac{2\omega_u^2}{\ln(\omega R/c)}, \tag{8.1a}$$

with $\omega_u = c/R$ given in Eq. (2.14b); R, the radius of the cylinder. A similar result was derived by Pendry [3] and coworkers. Their analysis is carried out for a wire radius less than the skin depth of the metal; the experiments [4] for the left-handed materials are performed for wires with radii larger than the skin depth. Eq. (8.1) provides for an extension of their analysis. Eq. (8.1) also clarifies the issue of damping. For frequencies from 1 to 10 GHz, the imaginary part of the dielectric constant of most metals is about a thousand times larger than the real part. When the skin depth is much less than the wire radius, the loss in the metal is only restricted near the surfaces of the wires and the effective damping is reduced. Indeed, the dielectric constant in Eq. (8.1) depends only on the wavelength and the wire radius, with no damping!

To illustrate the application of the multiple scattering method to composites we study a simple example of the photonic band structure of a two dimensional (2D) photonic crystals (PC) consisting of arrays of metallic or dielectric cylinders (wires) of radius R in an insulating matrix or arrays of insulating cylinders in a metallic matrix. There is much interest recently in left-handed materials and in plasmonics in this composite. Our calculation brings out several key ideas. (1) Even in the long wavelength limit when the separation between the wires a is less than the free space wavelength $\lambda = 2\pi/k_0$, sometimes more than one scattering resonances can be important. We find that the scattering phase shifts for both the s wave ($n = 0$ partial wave) and the p wave ($n = \pm 1$ partial waves) are of the same order of magnitude, $(k_0 R)^2$, and need to be considered. (2) The calculation shows how to make connection of the scattering of a single cylinder to the scattering from an array of them. The multiple scattering equations can be analytically solved in the long wavelength limit. This provides us with a physical feeling of the parameters involved in the calculation. (3) This calculation provides an understanding of *two* typical photonic branches that are often discussed, an "acoustic" mode with a frequency proportional to the wave vector with an effective dielectric constant $\langle \varepsilon \rangle$ and a magnetic susceptibility $\langle \mu \rangle$ and an "optic" mode with a gap. For *negative* dielectric constants and narrow cylinders, the "optic" mode corresponds to a flat band at frequencies close to the surface plasmon resonances, as has been previously discovered numerically. For negative magnetic susceptibilities, a "magnetic surface plasmon" band is found. One area of application of plasmonics is subwavelength transport of light. The crucial requirement for this application is that the wave vector perpendicular to the direction of motion is imaginary. This type of flat bands is intimately connected with different ways to fulfill the required conditions.

8.2 Scattering

We first briefly recapitulate the language of scattering theory. The scattering of the EM wave from a single structure can be summarized by the t matrix that is defined by a relationship between the scattered electric field and the incoming field:

$$\boldsymbol{E}_{\mathrm{sca}} = t\boldsymbol{E}_{in}. \tag{8.2}$$

For the well-known example of a sphere, its scattering properties as is first described by Mie are well known. The t matrix can be evaluated by the matching of the electromagnetic fields at the boundary.

The simplest example of the t matrix is that of the scattering from a planar interface. In that case, the scattered field in the forward (backward) direction is just the refracted field minus the incoming field (reflected field). When the electric field is parallel to the plane of incidence, the t matrix has been discussed in elementary physics textbooks and is given by [5]

$$t = [(\varepsilon'/\mu')k_z - (\varepsilon/\mu)k_z']/[(\varepsilon'/\mu')k_z + (\varepsilon/\mu)k_z']$$

where z is the direction of the surface normal, the unprime quantities are that of the incoming wave. The scattering becomes resonant when t becomes infinite and the denominator becomes zero. It is straightforward to show that the zero denominator condition leads to the equation

$$(\omega/c)^2 = k_x^2(\mu/\varepsilon + \mu'/\varepsilon'),$$

the same condition as that for a surface plasmon resonance. (Of course, in that case k_z and k_z' are both imaginary.) The infinite denominator is believed [6] to be the reason for the enhanced transmission through subwavelength holes in a metallic film in the infrared regime [1]. Enhanced transmission through sub-wavelength holes in metallic plates were observed in both infrared and microwave frequencies. In the microwave regime, the damping is higher by several orders of magnitude; the reason for the enhanced transmission is most likely different. Recent results by the HKUST group [7] show that the TE and the TM modes are coupled and the field is confined inside the holes. This suggests that the picture proposed by Pendry and coworkers [8] may need to be modified.

For scattering from a finite size object, quite often the t matrix is written in the form

$$t = 1/(1 + i\mathbf{C}). \tag{8.2a}$$

For scattering from spheres or from cylinders at perpendicular incidence, t is diagonal in angular momentum space. $\mathbf{C}$ is then just the cotangent of the scattering phase shift.

8.3 Effect of the Variation of the Current Along the Wire

For metallic wire structures, if the focus is on the variation of the current along the wire, this t matrix can be evaluated with the method developed in previous chapters as follows. From Eq. (5.10) we obtain the current induced in the metallic wire structure by an external field as

$$j = \mathbf{A}\mathbf{E}_{in} \tag{8.3}$$

where the matrix A is given by

$$A = \mathbf{X}^{-1} - \mathbf{Y}^{-1}(H_0\mathbf{X}^{-1})/\sum_h (H_0\mathbf{Y}^{-1})_h. \tag{8.3a}$$

From Maxwell's equations this current in turn generates the scattered field [see, for example, Eqs. (1.1) and (1.2)]

$$E_{\mathrm{sca}}(\vec{r}) = Gj$$

where

$$Gj = \frac{-i\omega}{c^2}(\vec{r}') \cdot (k^2 + \nabla\nabla)G(r - r', \omega)d^3\vec{r}'$$

$$G = \frac{\exp(ik|\vec{r} - \vec{r}'|)}{|\vec{r} - \vec{r}'|}.$$

Examples of this kind of calculation were discussed in Sect. 3.5. Combining with Eq. (8.3), we finally obtain

$$t = GA. \tag{8.4}$$

In the long wavelength limit the susceptibilities can be obtained from the above.

8.4 Variation Perpendicular to the Wire: The "Pseudopotential" Idea

There is another way to estimate the susceptibilities. To illustrate our idea we consider an infinitely long metallic cylinder of radius R with the EM wave coming in perpendicular to this cylinder. For this problem of cylindrical symmetry, the scattering states are characterized by an angular momentum index n. There are two scattering channels, the transverse electric (TE, B along cylinder axis) and the transverse magnetic (TM, E along cylinder axis) modes. For example, for the TM mode, the scattered and the incoming electric fields along the cylinder axis (the z axis) are given in cylindrical coordinates and in terms of Bessel functions by

$$E^{\text{sca}} = -\sum_n a_n H_n(k_o r)\exp(in\phi), \quad E^i = \sum_n p_n J_n(k_o r)\exp(in\phi) \tag{8.5a}$$

with constant coefficients a_n and p_n. The minus sign for the scattered wave was used historically and we follow this convention. For the TE mode, a corresponding expression for the magnetic field H along the cylinder axis can be written down. The total electric field is the sum of these two terms and it can be written as

$$E^{\text{tot}} = \sum_n i\cot\eta_n[J_n(k_o r) - \tan\eta_n N_n(k_o r)]\exp(in\phi)a_n \tag{8.5b}$$

in terms of the phase shift η_n defined by

$$a_n/p_n = 1/(1 + i\cot\eta_n). \tag{8.5c}$$

Equation (8.5c) for the phase shift may appear mysterious to the nonexpert. The expression for the total field in Eq. (8.5b) makes this identification clear. More precisely for large $x = k_o r$ $J_n(x)$, $N_n(x)$ are proportional to $\cos(x - n\pi/2 - \pi/4)$ and $\sin(x - n\pi/2 - \pi/4)$. Thus the nth component of the total electric field is proportional to $\cos(x - n\pi/2 - \pi/4 + \eta_n)$. This is what we expected of the phase shift.

The Mie scattering result from the cylinder, obtained from matching the tangential E and the H fields at the boundary, is well known [9]. The tangent of the phase shift η is given, for the E (TM) and H (TE) modes, respectively, by

$$\tan \eta_n^E = \frac{k_i J_n'(x) J_n(y) - k_o \varepsilon J_n(x) J_n'(y)}{k_i N_n'(x) J_n(y) - k_o \varepsilon N_n(x) J_n'(y)}$$

$$\tan \eta_n^H = \frac{k_i J_n'(x) J_n(y) - k_o \mu J_n(x) J_n'(y)}{k_i N_n'(x) J_n(y) - k_o \mu N_n(x) J_n'(y)} \tag{8.5}$$

The subscripts i, o refers to quantities inside and outside the cylinder, respectively. $k_j = k_0 (\mu_j \epsilon_j)^{1/2}$ for $j = o$ and i, $\epsilon = \epsilon_i / \epsilon_o$, $\mu = \mu_i / \mu_o$, and $x = k_o R$, $y = k_i R$. Eq. (8.5) is derived under the convention that the plane wave has a phase dependence of $-ikr$. If the convention is changed so that the sign of the phase is changed to $+ikr$, then the sign of the phase shift is also changed. We first focus on the s wave with $n = 0$. When the wavelength outside is much larger than the cylinder radius, $x = k_o R \ll 1$. We obtain $J_0(x) = 1$, $J_0'(x) = -x/2$, $N_0(x) = (2/\pi) \ln x$, and $N_0'(x) = 2/(\pi x)$ and hence

$$\tan \eta_0^E \approx -\frac{\pi x^2}{4} \frac{1 + 2\varepsilon J'/(yJ)}{1 - yJ' \ln x/(\mu J)}$$

$$\tan \eta_0^H = -\frac{\pi x^2}{4} \frac{1 + 2\mu J'/(yJ)}{1 - yJ' \ln x/(\varepsilon J)} . \tag{8.6}$$

where $J = J_0(y)$ and $J' = J_0'(y)$. If $y = k_i R$ is also small, we have

$$\tan \eta_0^E \approx -\frac{\pi}{4} x^2 [1 - \varepsilon], \quad \tan \eta_0^H \approx -\frac{\pi}{4} x^2 [1 - \mu].$$

As is expected, when $\varepsilon = 1$, there is no scattering and $\tan \eta_0^E = 0$. When $k_i R$ is not small, one can *define* effective susceptibilities so that the same phase shift is produced:

$$\tan \eta_0^E \approx -\frac{\pi}{4} x^2 [1 - \varepsilon_E'], \quad \tan \eta_0^H \approx -\frac{\pi}{4} x^2 [1 - \mu_H']$$

This is the "pseudopotential" idea that we mentioned. From Eq. (8.6) we obtain

$$\epsilon_E' = -\frac{\epsilon J'}{yJ} \frac{2 + x^2 \ln x}{1 - yJ' \ln x/(\mu J)},$$

$$\mu_H' = -\frac{\mu J'}{yJ} \frac{2 + x^2 \ln x}{1 - yJ' \ln x/(\epsilon J)} . \tag{8.7}$$

for the effective dielectric constant and magnetic susceptibility. A similar result can also be derived for chiral magnetic systems [10]. For metallic cylinders, the magnitude of the wave vector inside the metal, k_i, is of the order of the inverse skin depth of the metal. When the skin depth is much less than the radius of the cylinder, $y = k_i R \gg 1$, the second term in the denominator of the effective susceptibilities in

Eq. (8.7) is larger than the first term; assuming the outer region to be air with $\epsilon_o = \mu_o = 1$, we obtain an effective dielectric constant of a metallic form

$$\epsilon'_E = 1 - \frac{\omega'^2_p}{\omega^2} \tag{8.8}$$

where ω'_P is given in Eq. (8.1). A similar formula can also be derived using equation (8.3) and Eq. (2.26a). To obtain the dielectric constant of a composite, the simplest approximation is to multiply that for a single wire by an additional factor of the volume fraction of the wires. Except for a different log correction, formula (8.1) is the same as that of Pendry and coworker [3], when the volume fraction factor is taken into account.

Equation (8.7) encompasses other recent results for dielectric rods. If the second term of the denominator is much smaller than the first term, we recover recent results [11, 12], namely,

$$\epsilon'_E \approx -\frac{2J'\varepsilon}{Jk_iR}$$

This pseudopotential idea has been used in designing negatively refracting material with cylinders with a high dielectric constant ferroelectric [13] and magnetic cylinders [14]. Equation (8.7) extends these results to more general regions of the parameter space.

For μ'_H, the second term in the denominator is of the order of $(\mu_i/\epsilon_i)^{1/2}x$ and is usually smaller than the first term. We obtain

$$\mu'_H \approx -\frac{2J'\mu}{Jk_iR}.$$

We next investigate the phase shifts for the higher order partial waves. In the limit $x = k_oR \ll 1$

$$\tan \eta_n^E = -\frac{\pi(x/2)^{2n}}{(n-1)!\,n!} \frac{\mu_o - n\mu_i J_n/(yJ'_n)}{\mu_o + n\mu_i J_n/(yJ'_n)},$$

$$\tan \eta_n^H = -\frac{\pi(x/2)^{2n}}{(n-1)!\,n!} \frac{\epsilon_o - n\epsilon_i J_n/(yJ'_n)}{\epsilon_o + n\epsilon_i J_n/(yJ'_n)}.$$

Here $J_n = J_n(k_iR)$ and $J'_n = J'_n(k_iR)$. When $|k_iR| \ll 1$,

$$\tan \eta_n^E = -\frac{\pi(x/2)^{2n}}{(n-1)!\,n!} \frac{\mu_o - \mu_i}{\mu_o + \mu_i},$$

$$\tan \eta_n^H = -\frac{\pi(x/2)^{2n}}{(n-1)!\,n!} \frac{\epsilon_o - \epsilon_i}{\epsilon_o + \epsilon_i}.$$

Note that whereas the $n = 0$ E mode phase shift involves the dielectric constant, for $n > 0$, the magnetic susceptibility comes in. For "plasmonics", the frequency is close to the interface plasmon frequency so that $\epsilon = -1$. At this frequency $\eta_n^H = \pi/2$. Scattering resonances are exhibited for the TE modes *for all* $n \neq 0$. For $n = 1$ the requirement that the same scattering phase shift should be obtained even when $k_i R$ is not small provides for the equations determining the effective susceptibilities:

$$\tan \eta_1^E = -\frac{\pi x^2}{4} \frac{\mu_o - \mu_E'}{\mu_o + \mu_E'},$$

$$\tan \eta_1^H = -\frac{\pi x^2}{4} \frac{\epsilon_o - \epsilon_H'}{\epsilon_o + \epsilon_H'}.$$

From these we obtain the effective susceptibilities

$$\mu_E' = \frac{\mu_i J_1}{J_1' k_i R}; \quad \epsilon_H' = \frac{\epsilon_i J_1}{J_1' k_i R}. \tag{8.9}$$

Similar equations have also been obtained [11] from a coherent potential approximation. The results here provide a different interpretation of their results. With the current view, "plasmonics" phenomena can also be manifested for non-metallic rods if $\epsilon_H' + \epsilon_o = 1$ and the same scattering phase shift is produced.

We next examine possible generalization of the pseudopotential idea to scattering units other than cylinders (or spheres). The scattering information is contained in the T matrix which, in the angular momentum basis, can be written as $T = \sum |n\rangle T_{nm} \langle m|$. In the long wavelength limit, the T matrix becomes $T = \sum |n\rangle T_{nm}^0(\epsilon, \mu)\rangle m|$. Effective tensors ϵ, μ may be obtained if the equations $T_{nm} = T_{nm}^0(\epsilon, \mu)$ can be solved.

We next turn our attention to the photonic bands.

8.5 Photonic Band Structure

There are many ways to study the photonic band structure. The discussion in this book is particularly well suited to the multiple scattering method. Near a scattering resonance, it takes many plane waves to expand the fields inside the cylinder accurately. The multiple scattering method bypasses this problem. We recapitulate this next.

In the multiple scattering theory, the scattered field from a site i, can be represented by the coefficients of the basis functions a_i as in Eq. (8.5a). In the following, we shall not be very explicit and call this the incoming field. This incoming field is related to that for the *total* incoming field f_i by the scattering matrix of a single structure, t, in the form $a_i = t f_i$. The *total* incoming field is a sum of the *external* incoming field p_i and the scattered waves from all the other scatterers, which can be written as: $f_i = \sum [j \neq i - S_{ij} b_j + p_i]$. Here, S_{ij} is a "structure factor" that transforms the outgoing wave from site j into an incoming wave at site i. The minus sign in front of S was used historically and we follow this convention. The structure factor is determined from the geometry of the system.

Examples of these for different lattices can be easily found in the literature. Usually these are evaluated with the Ewald sum technique. For finite size systems, these can be evaluated directly by brute force.

The scattering amplitude a from a collection of the structures is related to the amplitude p_i of the external wave by the full T matrix:

$$a_i = Tp_i.$$

From the above reasoning, we find that the inverse T matrix of the total collection of structures is a sum of the single cylinder inverse t matrix and the structure factor S:

$$T^{-1} = t^{-1} + S \qquad (8.10)$$

The photonic band frequency is determined from the condition

$$T^{-1} = 0. \qquad (8.10a)$$

Again, the determinant of T^{-1} is usually very large and it is numerically difficult to directly determine the frequencies so that this determinant is zero. Eq. (8.10a) can be solved numerically with the method described in Chap. 4, Sect. 8.2. The matrix M now corresponds to the matrix T^{-1}.

For periodic systems, T is diagonal in the wave-vector space. For finite size systems, this is no longer true. The whole matrix can be evaluated directly numerically. To illustrate this method we show how the long wavelength limit is recovered for a collection of infinitely long cylinders with the EM wave vector in the plane perpendicular to them [15]. This is a very instructive calculation, as it shows us how this method can give us the dispersion.

In 2D, the scattered wave is characterized by a single index, the angular momentum n. We denote the amplitude of the scattered wave by a_n. We get the sum of the scattered waves from the sites R given by

$$S''(n) = \sum_{R \neq 0} \exp(ip \cdot R) H_n(k|r - R|) \exp(in\theta_1) a_n.$$

The Gegenbauer expansion [16] allows us to express this in terms of the incoming wave at the origin: For $R + r_1 = r$, we have

$$H_n(kr_1) e^{in\theta_1} = \sum_m H_m(kR) A'^{-im\theta_R}$$

where $A'(n + m) = J_{n+m}(kr) e^{i(m+n)\theta}$ is the $m + n$ component of the incoming wave at the origin. Substituting this into the expression for S'', we obtain $S''(n) = \sum_m S(m) A'(n + m) a_n$ where the structure factor is given by

$$S(m) = \sum_{R \neq 0} H_m(kR) \exp(-im\theta_R) \exp(ik \cdot R). \qquad (8.11)$$

With the convention that the plane wave has a phase dependence of $-ikr$ [$+ ikr$], it is $H_m^{(2)}$ [$H_m^{(1)}$] that appears in Eq. (8.11). The sum of the scattered waves from all the *other* sites becomes an incoming wave at the origin with the amplitude $p_n = {}_n a_{n'} S(n - n')$. Note that S does *not* include the wave from the origin; thus the sum in Eq. (8.11) does not include the term at $R = 0$. The outgoing scattered wave at the origin is related to the incoming wave by the t matrix: $a_n = t_n p_n$. With the definition of p_n, we arrive at the equation (8.10a), which can be written explicitly as:

$$\det[S(n - n') - \delta_{nn'}/t_n] = 0.$$

Here $\delta_{nn'}$ denotes the Kronecker δ -symbol. Since the t matrix is related to the phase shift by Eq. (8.2a):

$$t_n = 1/(1 + i \cot \eta_n), \tag{8.12}$$

we obtain the multiple scattering equation

$$\det[A(n - n') - \delta_{nn'} \cot \eta_n] = 0. \tag{8.13}$$

Here $A(n) = [S(n) - \delta_{n0}]/i$.

In the long wavelength limit, one can approximate the sum for S by an integral, which can then be analytically evaluated. The details are explained in the Appendix A in this chapter. The "structure factor" becomes

$$A(n) \approx \frac{4\, i^n e^{in\phi_k} k^n}{k_o^n (k^2 - k_o^2) a^2} \tag{8.13a}$$

This result provides us with an explicit expression and a physical feeling for S, which is basically a Green's function. As expected, S possesses a pole on the light cone, when the frequency is equal to the speed of light times the wave vector. This is useful for understanding the numerical results for S away from the long wavelength limit.

As is discussed above, if the wavelength outside the cylinder is long and $k_o R \ll 1$, $\tan \eta_n \propto (k_o R)^{2n}$ for $n \neq 0$, $\tan \eta_0 \propto (k_o R)^2$. The phase shifts for the s and the p waves are of the same order of magnitude, $(k_o R)^2$, and need to be considered. This is different from the scattering from spheres, where only s wave scattering is important. When only the s and the p wave components are kept, the multiple scattering equation (8.13) reduces to

$$HE = 0$$

where

$$H = \begin{bmatrix} A(0) - \cot \eta_1 & A(1) & A(2) \\ A(1)* & A(0) - \cot \eta_0 & A(1) \\ A(2)* & A(1)* & A(0) - \cot \eta_1 \end{bmatrix}$$

There are two classes of solutions, with either $E_1 = E_{-1}^* = |E_1|e^{i\phi_k}$, $E_0 = E_0^*$ or $E_1 = -E_{-1}^* = i|E_1|e^{i\phi_k}$, $E_0 = 0$. We get two possible eigenvalue equations. The first one is given by

$$[A(0) - \cot\eta_1 + |A(2)|][A(0) - \cot\eta_0] - 2|A(1)|^2 = 0. \tag{8.14}$$

For the second case, we get

$$A(0) - \cot\eta_1 + |A(2)| = 0. \tag{8.15}$$

As we show below, the first mode corresponds to an "acoustic" branch with a frequency proportional to the wave vector, enabling an effective medium description for the system; the second mode corresponds to a band with a gap. For negative susceptibilities, this corresponds to a flat band at frequencies close to the surface plasmon resonances, as has been previously found numerically [2]. This second branch is closely related to the issues of subwavelength transport of light.

We discuss the acoustic branch first. Substituting in the expressions for the phase shifts and the structure factor into Eq. (24) and after some algebra (See Appendix B in this chapter), we obtain

$$k^2 = k_0^2 \langle \epsilon \rangle \langle \mu \rangle \tag{8.16}$$

where

$$\begin{aligned}
\langle \epsilon \rangle &= (1 - f)\epsilon_o + f\epsilon_i', \\
\langle \mu \rangle &= \mu_o \frac{\mu_i'(1 + f) + \mu_o(1 - f)}{\mu_i'(1 - f) + \mu_o(1 + f)} .
\end{aligned} \tag{8.16a}$$

with ϵ_i' and μ_i' denoting the effective susceptibilities of the cylinders.

In the static (zero wave vector and frequency) limit, for the case with the E field along the axis, Eq. (8.16) reduces to $\langle \epsilon \rangle = (1 - f)\epsilon_o + f\epsilon_i$, implying that the average dielectric constant is just the arithmetic mean of the dielectric constants of the components, as is well known [17].

In multilayer systems, a similar result is obtained [18]. In that case, the effective μ is the harmonic mean of the components while the effective dielectric constant is still the arithmetic mean of that of its components.

We discuss next the "optic" mode. Substituting in the expressions for the phase shifts and the structure factor $A(n)$, the equation for the second optic mode becomes

$$\frac{2(k_o a)^2}{\pi} \ln\frac{k_o a}{2\sqrt{\pi}} = 4 - 4f^{-1}\frac{\mu_o + \mu_E'}{\mu_E' - \mu_o} + O(k^2)$$

for the E mode and

$$\frac{2(k_o a)^2}{\pi} \ln\frac{k_o a}{2\sqrt{\pi}} = 4 - 4f^{-1}\frac{\epsilon_o + \epsilon_H'}{\epsilon_H' - \epsilon_o} + O(k^2)$$

for the H mode. When k approaches zero, k_o is not zero! Let us illustrate the physics by looking at the H mode. The limit of small f is particularly interesting. In that limit, the frequency is determined by the condition that $\epsilon_o + \epsilon'_H = 0$ where ϵ'_H is given in Eq. (8.9).

For metallic cylinders with their radii less than the skin depths, $k_i R \ll 1$, $\epsilon'_H = \epsilon_i = 1 - \omega_p^2/\omega^2$ where ω_p is the plasma frequency. $\epsilon_o + \epsilon_i = 0$ when ω is equal to the interface plasmon resonance, $\omega_{sp} = \omega_p/(1 + \epsilon_o)^{1/2}$, For small k, from the above equation, we see that $\omega(k) = \omega(k = 0) + O(fk)$. If f is small, the dispersion is weak and the bands are flat. This flat band has been observed numerically previously [2]. The present calculation provides for a more direct analytic demonstration of this result. If $k_i R$ is not small, Eq. (8.9) suggests that even with insulating cylinders, flat "plasmonic" photonic bands can still be obtained if the following condition is satisfied,

$$\frac{\epsilon_i J_1}{J'_1 k_i R} = -\epsilon_o.$$

Let us next look at the E mode, the condition becomes $\mu'_E/\mu_o = -1$. We call this the "magnetic surface plasmon" mode. Although a lot of interest in plasmonics has been focused on the condition $\epsilon'_H/\epsilon_o = -1$, the counterpart condition on μ have not been much discussed. Magnetic surface plasmons are "one-way" because time reversal symmetry is broken in magnetic systems.

There is another way to think of this type of solutions. When the susceptibilities of the metal are negative, $\epsilon_o + \epsilon_m$ can become zero and $\tan \eta = \infty$. The scattering can go though resonances due to the interface plasmon. This can lead to flat photonic bands, as has been observed in previous numerical calculations. In general, the more rapidly the phase shift varying, the flatter the band.

In conclusion, in this section we use a pseudopotential idea to derive effective susceptibilities of cylinders so as to mimic the scattering phase shifts of the system. We calculate analytically the long wavelength limit photonic band dispersion in a 2D photonic crystal and explicitly demonstrate the flat "surface plasmon" photonic bands with small group velocity v_g, which are implicitly exploited in the study of plasmonics. The spatial extent of a wavepacket Δx is of the order of $v_g/\Delta\omega$. For subwavelength localization $\Delta x \ll \lambda$ where λ is the wavelength. We thus obtain the condition $\Delta\omega \gg v_g/\lambda$. Because v_g is small, the spread in frequency $\Delta\omega$ remains small!

At the surface plasmon frequency, the phase shift is equal to $\pi/2$ in the limit when the cylinder radius is small. Our pseudopotential idea suggests that "plasmonics" effect need not be restricted to metallic systems at the surface plasmon frequency. Other systems with the same resonance phase shift will lead to similar photonic bands to provide subwavelength transmission and thus can serve as alternative candidates [19].

Appendix A

In this appendix we evaluate the structure factor analytically in the long wavelength limit. We change the variable R to $x = k_o R$ and get

$$S(0) = \sum_{x \neq 0} (\Delta x)^2 \exp(ik \cdot x/k_o) H_0(x)/(k_o a)^2$$

In the long wavelength limit, Δx becomes small. We approximate this sum by an integral and get

$$S(0) = \int_{x_i}^{\infty} d^2 x \exp(ik \cdot x/k_o) H(x)/(k_o a)^2$$

We pick the lower limit so that the empty area remains the same $[\pi x_i^2 = (k_o a)^2]$. Since $e^{ia\cos(\theta)} = \sum_m i^m J_m(a) e^{im\theta}$, only the m = 0 term remains in the integral. We get

$$S(0) = 2\pi \int_{x_i}^{\infty} x dx J_0(kx/k_o) H(x)/(k_o a)^2$$

The radial integral is a standard integral that occurs in the computation of the ortho-normal properties of the Bessel functions and can be analytically done. [We assume that k_0 has a small imaginary part so that the upper limit contribution can be set to zero.] We get,

$$S(0) \sim -2\pi x \left[k J_0'(xk/k_o) H_0(x) - k_o J_0(xk/k_o) H_0'(x) \right] / \left[(k_o a)^2 (k^2 - k_o^2) \right] \Big|_{xi}$$

Using the small argument expansions for the Bessel and Hankel functions in the limit $x \ll 1$: $J_0(x) \approx [1 - (x/2)^2]$, $J'(x) \approx -x/2 + O(x^3)$, $N_0'(x) = 2\ln(x/2)[1 - x^2/4]/\pi$, $N_{0'}(x) = 2[(1 - x^2/4)/x - x\ln(x/2)/2]/\pi$
we obtain

$$S(0) \sim 1 + \{4i[k_o^2 + (x_i^2/2)(k^2 - k_o^2)]\ln(x_i/2)\}/\left[(k_o a)^2(k^2 - k_o^2)\right],$$

Substituting in the value of x_i, we obtain

$$A(0) = [S(0) - 1]/i \sim \{2[2\pi k_o^2 + (k_o a)^2(k^2 - k_o^2)]\ln(k_o a/2\pi^{1/2})\}/[\pi(k_o a)^2(k^2 - k_o^2)]$$

Similarly, for n > 0,

$$J_n(x) \approx (x/2)^n/n![1 - (x/2)^2/(n + 1)]$$
$$J'(x) \approx (n/x)(x/2)^n/n![1 - (n + 2)(x/2)^2/(n(n + 1))]$$
$$N_n(x) = -(2/x)^n(n - 1)![1 + n(x/2)^2]/\pi.$$
$$S(n) \sim -2\pi i^n(k_x + ik_y)^n x \left[k J_n'(xk/k_o) H_n(x) - k_o J_n(xk/k_o) H_n'(x) \right] / \left[(k_o a)^2 (k^2 - k_o^2) \right] \Big|_{xi}$$

Substituting in the value of x_i, we finally get

$$A(n) \approx \frac{4\,i^n e^{in\phi_k} k^n}{k_o^n (k^2 - k_o^2) a^2}$$

Appendix B

In this appendix we derive Eq. (8.16) from Eq. (8.14). Define $u = k_o^2/(k^2 - k_o^2)$, then $A(0) \approx 4u/(k_o a)^2$, $\exp(-in\phi)A(n) \approx \alpha_n A(0)$ with $\alpha_n = (ik/k_o)^n$. Noting that the phase shifts are $\tan\eta_0 \approx \pi(k_o R)^2/(4c_0)$ and $\tan\eta_1 \approx \pi(k_o R)^2/(4c_1)$, where $c_0 \approx -\epsilon_o/(\epsilon_o - \epsilon_c)$, $c_1^{-1} \approx -(\mu_o - \mu_c)/(\mu_o + \mu_c)$, with ϵ_c and μ_c being the effective susceptibilities of cylinder, Eq. (8.14) can be written as $u^2[1 + |\alpha_2| - 2|\alpha_1|^2] - (u/f)[c_1 + 1 + |\alpha_2|]c_0 + c_1 c_0/f^2 = 0$ where $f = \pi R^2/a^2$ is the volume fraction of the cylinders. Now $u[1 + |\alpha_2| - 2|\alpha_1|^2] = -1$. We thus get

$$uf[-c_1 - f + (1 + |\alpha_2|)c_0] + c_1 c_0 = 0$$

Substituting in the value of u, this can be written as

$$f[-c_1 - f + (1 + k^2/k_o^2)c_0] = -c_1 c_0 (k^2/k_o^2 - 1) = 0$$

Inserting the definitions of c_0 and c_1 yields equation (8.16):

$$k^2 = k_0^2 \langle \epsilon \rangle \langle \mu \rangle$$

where

$$\langle \epsilon \rangle = (1 - f)\epsilon_o + f\epsilon_i',$$
$$\langle \mu \rangle = \mu_o \frac{\mu_i'(1 + f) + \mu_o(1 - f)}{\mu_i'(1 - f) + \mu_o(1 + f)}.$$

References

1. T.W. Ebbesen, H.J. Lezec, H.F. Ghaemi, T. Thio, P.A. Wolff, Nature **391**, 667 (1998)
2. Plasmonics: Fundamentals and Applications. Springer; A. V. Zayats, I. I. Smolyaninov, A. A. Maradudin, "Nano-optics of surface plasmon polaritons", Phys. Rep., 131 (2005) Harry A. Atwater (2007). The Promise of Plasmonics. In Scientific American, April 2007 v.296 n.4, pg.56-63
3. J. B. Pendry, A. J. Holden, D. J. Robbins and W. J. Stewart, J. Phys. Conds. Matt. 4785 (1998)
4. R. A. Shelby, D. R. Smith, S. C. Nemat-Nasser, S. Schultz, Appl. Phys. Lett. 489 (2001)

5. R. P. Feymann, R. B. Leeighton M. Sands, *Lectures on Physics*, vol. 2, eq. (33.53), published by (Addison-Wesley Reading, Massachusetts) (1964)
6. L. Martin-Moreno et al., Phys. Rev. Lett. **86**, 1114 (2001)
7. B. Hou, Z. H. Hang, W. Wen C. T. Chan, P. Sheng, Appl. phys. lett. **89**, 131917 (2006)
8. J. B. Pendry, L. Martín-Moreno, F. J. García-Vidal, Science **305**, 847 (2004)
9. H. C. van de Hulst, Light scattering by small particles, (Dover NY 1957)
10. J. Jin, S. Liu, Z. Lin, S. T. Chui, Phys. Rev. B **80**, 115101 (2009)
11. Y. Wu, J. S. Li, Z. Q. Zhang, C. T. Chan, Phys. Rev. B, 085111 (2006)
12. X. H. Hu, C. T. Chan, J. Zi, M. Li, and K. M. Ho, Phys. Rev. Lett., 223901 (2006)
13. L. Peng, L. X. Ran, H. S. Chen, H. F. Zhang, J. A. Kong, and T. M. Grzegorczyk, Phys. Rev. Lett. 157403 (2007)
14. S. Liu, W. Chen, J. Du, Z. Lin, S. T. Chui C. T. Chan, Phys. Rev. Lett. **101**, 157407, (2008)
15. S. T. Chui, Z. F. Lin, Phys. Rev. E **78**, 065601(R) (2008)
16. Electromagnetism, by Stratton, eq. 6.1.13 p. 374
17. M. Born E. Wolf, 7th edn. (Cambridge University Press, 1999), p. 837
18. S. T. Chui, C. T. Chan, Z. F. Lin, Jour. Phys. Conds. Matt. L89 (2006)
19. Du JJ, Lin ZF, Chui ST, et al., Phys. Rev. Lett. **106**, 203903 (2011) and references therein

Chapter 9
Electromagnetic Waves in Wire Composites II: Anisotropic, Off-Diagonal Magnetoelectric Wire Composites

9.1 Introduction

In this last chapter we continue to consider the situation where there is a collection of metallic wire structures and examine the propagation of EM waves in them. Metallic wire structures are often magnetoelectric [1–3] and lead to many unexpected properties. This chapter focuses on the physical properties of a collection of such materials. The magnetoelectric property is also called bi(an)isotropic in the electrical engineering community. We first remind the reader what this means.

The long wavelength behavior of an EM wave in a material is governed by its electric and magnetic susceptibilities. In ordinary materials, the magnetization is caused by an external magnetic field whereas the electric polarization is caused by an external electric field. For magnetoelectric material, an external electric (magnetic) field can cause a magnetic (electric) polarization, that is,

$$\mathbf{M} = \hat{\alpha}_m \mathbf{B} + \hat{\alpha} \mathbf{E}, \quad \mathbf{P} = \hat{\beta}_e \mathbf{E} + \hat{\beta} \mathbf{B}. \tag{9.1}$$

where $\hat{\beta} = -\hat{\alpha}^T$, (superscript T denotes the transpose) generally.

Examples of the evaluation of these susceptibilities were given in Sect. 4.5.

Sometimes it is convenient to express the moments in terms of H instead of B. From Eq. (9.1) we get

$$\mathbf{M} = \hat{\alpha}'_m \mathbf{H} + \hat{\alpha}' \mathbf{E}, \quad \mathbf{P} = \hat{\beta}'_e \mathbf{E} + \hat{\beta}' \mathbf{H}.$$

where

$$\hat{\alpha}'_m = (1 - 4\pi\hat{\alpha}_m)^{-1}\hat{\alpha}_m, \ \hat{\alpha}' = (1 - 4\pi\hat{\alpha}_m)^{-1}\hat{\alpha}, \ \hat{\beta}' = \hat{\beta}(1 + 4\pi\hat{\alpha}'_m),$$

$$\hat{\beta}'_e = \hat{\beta}_e + 4\pi\hat{\beta}\hat{\alpha}'.$$

Materials that are both ferromagnetic and ferroelectric at the same time (multiferroic) exhibit these types of phenomena. Usually the magnetoelectric coefficients are small. Interests in these types of materials have recently revived (for some

S. T. Chui and L. Zhou, *Electromagnetic Behaviour of Metallic Wire Structures*,
DOI: 10.1007/978-1-4471-4159-4_9, © Springer-Verlag London 2013

recent references, see, for example, [4, 5]) due to improved sophistication in creating multiphase nanostructure material. Fresh from the experience gained in making perovskite-type high temperature superconductors, new classes of multiferroic composite materials with larger magnetoelectric coefficients was recently made.

As is discussed in the previous chapters the split ring and the helix system are also bi(an)isotropic and magnetoelectric. Whereas the magnetoelectric effects in multiferroic materials decrease drastically above the spin wave frequency, the magnetoelectric effect in the split rings can persist up to much higher frequencies with wavelengths of the order of their sizes. A new aspect of the split ring structure is that it is anisotropic; furthermore, the magnetoelectric coefficients are often off-diagonal. Regardless of the form of the magnetoelectric coefficients, the dispersion can always be calculated analytically in the long wavelength limit. Because of the constraint $\nabla \cdot B = 0$, there are only two degrees of freedom. Thus the eigenvalue equation is always quadratic and not cubic. The algebra can be complicated. The equations involved can be written in matrix form. The algebra can then be handled by a symbol manipulation routine, which has the capability of handling matrix calculation. This chapter provides some examples of how this can be done.

There has been much work done on the propagation of EM waves through magnetoelectric materials where the magnetoelectric coefficients are *isotropic and diagonal* [6]. When another component with complementary resonant behavior is added, the total composite can be negative refracting [7]. However, very little work was done to study the physics in the anisotropic off-diagonal systems such as the split ring system. The *anisotropic* magnetoelectric effect of the split ring structure is interesting in its own right. As a result of these, many novel things can happen [8]. We describe some examples of these effects next.

(a) *Longitudinal elliptic polarization.* The electric field of electromagnetic waves can be polarized in different ways. For elliptic polarization the electric fields along two orthogonal directions are 90° out of phase. In nearly all materials, such polarization is transverse in that the electric field is perpendicular to the wave vector. Very little is known whether there exist materials such that the elliptic polarization is *longitudinal* wherein the field rotates between a direction along the wave vector **k** and another direction perpendicular to it. In addition the consequence of this longitudinal elliptic polarization has not been explored.

Longitudinal elliptic polarization can be exhibited for EM waves propagating in a collection of metallic wire structures. To illustrate we consider the example of the propagation of plane electromagnetic waves through different split ring systems consisting of arrays of rings of different orientations. We replace the system by an effective medium with different average susceptibilities and find different modes of propagation with different velocities. For isotropic systems, the speed of light is inversely proportional to the square root of $\mu\epsilon$. In the present case, for one of the modes the square of the velocity has a positive real part even when $\mu\epsilon$ is negative; in addition, the electric field becomes longitudinally elliptically polarized.

(b) *Transverse Poynting vector.* For ordinary materials, the direction of energy flow, given by the Poynting vector $E \times H$, is along the direction of the wave vector. For the left-handed material, the focus is on the frequency region where the direction of the Poynting vector is opposite to the wave vector. For a longitudinal elliptically polarized wave, the cross product of a component of E along the wave vector k and a component of H perpendicular to k produces a component of the Poynting vector *perpendicular* to the wave vector even though the group velocity is *parallel* to the wave vector.

The above are some examples of new phenomena that can be exhibited. This chapter is devoted to understanding possible magnetoelectric properties of different simple structures such as split rings and helixes in the long wavelength limit. Helixes are chiral and possess diagonal magnetoelectric coefficients whereas the magnetoelectric coefficients of the split rings are off-diagonal. In the short wavelength limit the problem can be addressed by the multiple scattering methods as is described in the previous chapter.

9.2 Propagation of Plane Waves

We first derive the equation determining the dispersion relation of a plane wave through a homogeneous magnetoelectric material. We shall start with Maxwell's equation

$$\nabla \times H = 4\pi J/c + \partial_t D/c.$$

Now $H = B - 4\pi M$, $D = E + 4\pi P$. Substituting in the magnetoelectric coefficients from Eq. (9.1), this equation becomes (assuming that the current density $J = 0$)

$$\nabla \times (\mu^{-1}B - 4\pi\alpha E) = \partial_t(\epsilon E + 4\pi\beta B)/c. \tag{9.2}$$

where $\hat{\mu}^{-1} = 1 - 4\pi\hat{\alpha}_m$, $\hat{\epsilon} = 1 + 4\pi\hat{\beta}_e$. For our problem, μ and ϵ are diagonal.

Sometimes it is convenient to express quantities in terms of H instead of B, we get

$$\nabla \times H = \partial_t(\epsilon E + 4\pi\beta' H)/c.$$

We look for plane wave solutions proportional to $\exp -i(\mathbf{k} \cdot \mathbf{r} - \omega t)$. Equation (9.2) becomes

$$k \times (\mu^{-1}B - 4\pi\alpha E) = -\omega(\epsilon E + 4\pi\beta B)/c. \tag{9.3}$$

In terms of H, we get

$$k \times H = -\omega(\epsilon E + 4\pi\beta' H)/c. \tag{9.3a}$$

From Eq. (9.3), we obtain a relationship between E and B as

$$B = [(k \times \mu^{-1} + \omega 4\pi\beta/c)]^{-1}(-\omega\epsilon/c + k \times 4\pi\alpha)E.$$

The corresponding relationship between E and H is

$$H = -[(k \times +\omega 4\pi\beta'/c)]^{-1}\omega\epsilon E/c.$$

The coupling with an external field (reflection and refraction) involves the EM wave passing through interfaces. At interfaces, the tangential components of E and H are continuous. A quantity that characterizes the reflectivity is the impedance. The above equations involve vectors and in general are quite complicated. For an order of magnitude, one can extract an effective interface impedance ($\sim H/E$):

$$Z \sim - [(k \times +\omega 4\pi\beta'/c)]^{-1}\omega\epsilon/c.$$

This is one quantity we shall focus on.

From the other Maxwell's equation $\nabla \times E + \partial_t B/c = 0$, Eq. (9.2) becomes

$$\nabla \times (\mu^{-1}B - 4\pi\alpha E) = \partial_t \epsilon E/c - 4\pi\beta\nabla \times E.$$

Taking the time derivative of this equation and dividing by c, we obtain

$$-\nabla \times \mu^{-1}\nabla \times \mathbf{E} - 4\pi\nabla \times \hat{\alpha}\partial_t\mathbf{E}/c = \partial_t^2 \epsilon\mathbf{E}/c^2 - 4\pi\hat{\beta}\nabla \times \partial_t\mathbf{E}/c.$$

We look for plane wave solutions to this equation. In wave vector-frequency space, this equation becomes

$$\mathbf{k} \times \hat{\mu}^{-1}\mathbf{k} \times \mathbf{E} - 4\pi\omega\mathbf{QE}/c = -\omega^2\hat{\epsilon}\mathbf{E}/c^2 \tag{9.3}$$

where

$$\mathbf{Q} = \mathbf{k} \times \hat{\alpha} - \hat{\beta}\mathbf{k} \times . \tag{9.3a}$$

Equation (9.3) is a central equation in this chapter. The dispersion relation of a plane wave is obtained from solving this eigenvalue equation. Different wire structures give rise to different magnetoelectric coefficients and thus different eigenvalue equations.

In the usual isotropic limit when $\hat{\alpha} = -\hat{\beta}$ are diagonal, Eq. (9.3) is easily solved. We obtain the relation

$$\omega^2\varepsilon/c^2 = k^2/\mu \pm 4\pi\omega\alpha\beta k/c. \tag{9.3b}$$

First consider the case with $\hat{\alpha} = 0$. In situations when the composite contains additional components so that the dielectric constant exhibits a resonance that is of the form $\epsilon = a + b/(\omega^2 - \omega_0^2)$ for some constants a, b. This dielectric constant becomes zero at a frequency $\omega_0' = (\omega_0^2 - b/a)^{1/2}$ so that $\epsilon \sim e(\omega - \omega_0')$ near this frequency. Substituting this form of the dielectric constant into Eq. (9.3b) we find a photonic branch with dispersion given by $\omega = \omega_0' + c^2k^2/(\mu e\omega_0'^2)$ in the small k limit. This branch is similar to the optical branch we discuss in Sect. 8.5. In the presence of the magnetoelectric coefficient, we obtain the dispersion $\omega = \omega_0' + [k^2/\mu \pm 4\pi\omega_0'\alpha\beta k/c]c^2/(\omega_0'^2 e)$. Two photonic branches with phase

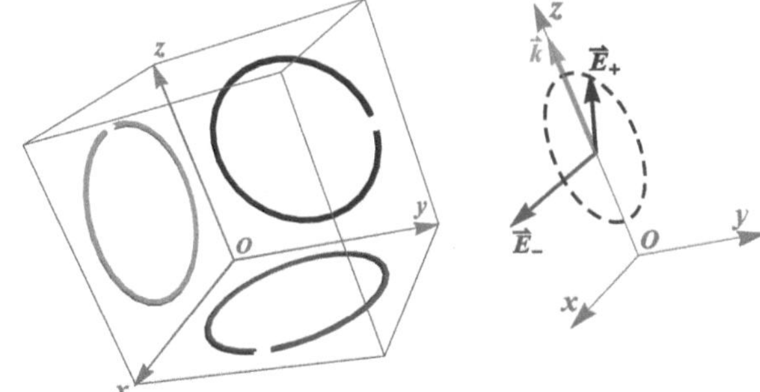

Fig. 9.1 *Left*: Illustration of the orientations of the three possible types of rings with the three cuts; *Right*: Illustration of the longitudinal elliptic polarization (E_+) and the transverse polarization (E_-) of the two normal modes found (from Ref. [8])

velocities $v = \pm 4\pi\alpha\beta c/(e\omega_0')$ are obtained. One of these phase velocities is negative and cause negative refraction, as is discussed by Pendry [7].

In this chapter we shall give examples of the solution of Eq. (9.3) for different systems when the magnetoelectric coefficients are anisotropic and off-diagonal. For analytic solutions of Eq. (9.3), we find it convenient to write it in matrix form in terms of the three components of the electric field. This makes it possible to carry out the algebraic computation with a symbol manipulation program, which has built-in capability of carrying out matrix algebra. A matrix that is often used is the cross product with the wave vector, which can be written as:

$$k\times = \begin{bmatrix} 0 & -k_z & k_y \\ k_z & 0 & -k_x \\ -k_y & k_x & 0 \end{bmatrix}.$$

9.2.1 One Type of Split Rings

We shall discuss split ring systems with up to three orientations, as illustrated in Fig. 9.1. We first consider the case where all the rings are oriented parallel to each other and in the xy plane. The anisotropic off-diagonal susceptibilities become:

$$\hat{\alpha} = \alpha_0 \begin{bmatrix} 0 & 0 & 0 \\ 0 & 0 & 0 \\ 0 & -1 & 0 \end{bmatrix}$$

$$\hat{\beta} = \alpha_0 \begin{bmatrix} 0 & 0 & 0 \\ 0 & 0 & 1 \\ 0 & 0 & 0 \end{bmatrix}.$$

The magnitude of α_0 corresponds to the quantity α_{yz}^{em} discussed in Eq. (3.5) for a split ring and χ_{zy} in (4.16) for a helix. For the special case with the wave vector along x and E along y, we get

$$(k_x \mu_{zz}^{-1} - 4\pi\beta_{yz}k_0)B_z = (4\pi\alpha_{zy}k_x + k_0\varepsilon_{yy})E_y. \tag{9.4}$$

The corresponding impedance is

$$Z_{zy} = (4\pi\alpha_{zy}k_x + k_0\epsilon_{yy})/(k_x - 4\pi\beta_{yz}\mu_{zz}k_0) - 4\pi\alpha_{zy}. \tag{9.5}$$

This quantity will be useful in the study of reflection and refraction of this composite.

In general

$$\mathbf{k} \times \alpha = \alpha_0 \begin{bmatrix} 0 & -k_y & 0 \\ 0 & k_x & 0 \\ 0 & 0 & 0 \end{bmatrix}$$

$$\hat{\beta}\mathbf{k}\times = -\alpha_0 \begin{bmatrix} 0 & 0 & 0 \\ -k_y & k_x & 0 \\ 0 & 0 & 0 \end{bmatrix}$$

Our eigenvalue Eq. (9.3) becomes

$$(\mu_{zz}^{-1} - 1)(k_y\mathbf{e}_x - k_x\mathbf{e}_y)(E_yk_x - E_xk_y) + (\mathbf{k}\mathbf{k}\cdot -k^2)\mathbf{E} - 4\pi\omega(\mathbf{e}_xE_y - \mathbf{e}_yE_x)$$
$$k_y\alpha_{zy}/c = -\omega^2\epsilon\mathbf{E}/c^2.$$

$$(\mu_{zz}^{-1} - 1)(k_y\mathbf{e}_x(E_yk_x - E_xk_y) - k_x\mathbf{e}_y(E_yk_x - E_xk_y)) + (\mathbf{k}\mathbf{k}\cdot -k^2)\mathbf{E} - k_0(\mathbf{e}_xE_y - \mathbf{e}_yE_x)$$
$$k_y\alpha' = -k_0^2\mathbf{E}.$$

where $k_0 = (\epsilon)^{1/2}\omega/c$, $\alpha' = 4\pi\alpha\sqrt{1/\epsilon}$, $\beta' = 4\pi\beta\sqrt{1/\epsilon}$.

$$(\mu_{zz}^{-1} - 1)(k_y\mathbf{e}_x(E_yk_x - E_xk_y) - k_x\mathbf{e}_y(E_yk_x - E_xk_y)) + (\mathbf{k}\mathbf{k}\cdot -k^2)\mathbf{E} - k_0(\mathbf{e}_xE_y - \mathbf{e}_yE_x)$$
$$k_y\alpha' = -k_0^2\mathbf{E}.$$

This can be written in matrix form as $\mathbf{\Omega E} = 0$ where

$$\hat{\Omega} = \begin{bmatrix} -k_y^2\mu^{-1} - k_z^2 + k_0^2 & \mu^{-1}k_xk_y - k_0k_y\alpha' & k_xk_z \\ \mu^{-1}k_yk_x + k_0\alpha'k_y & -k_x^2\mu^{-1} - k_z^2 + k_0^2 & k_yk_z \\ k_zk_x & k_yk_z & k_z^2 - k^2 + k_0^2 \end{bmatrix}$$

For $k_z = 0$, only the x and y components of E are coupled with each other. We can set $E_z = 0$, diagonalize a 2×2 matrix and get the eigenfrequencies: $k_0^2 = k_p^2$,

$k_0^2 = k_p^2/\mu - k_y^2 a^2$. Although the electric field is in the xy plane, from Eq. (9.4) the z component of the magnetic field is not equal to zero. Thus Z_{zy} (Eq. 9.5), which measures the energy flow in a direction perpendicular to the wave vector, is not zero. Thus even in this simple case, the phenomena we mentioned in the introduction are already manifested.

For a nonzero k_z, using a symbol manipulation program, we obtain

$$\det(\Omega)/k_0^2 = k_0^4 - [k_p^2(1 + 1/\mu) - k_y^2 a^2 + 2\,k_z^2]k_0^2 + [-k_y^2 a^2 k_p^2 + k_p^4/\mu + k_z^4 \\ + k_p^2 k_z^2(1 + 1/\mu)] \tag{9.6}$$

where $k_p^2 = k_x^2 + k_y^2$. As discussed above, because of the constraint $\nabla \cdot B = 0$ there are only two degrees (not three) of freedom. The eigenvalue equation is just a quadratic (and not cubic) equation in k_0^2. For the special case of $k_z = 0$, this equation reduces to $\det(\Omega)/k_0^2 = k_0^4 - [k_p^2(1 + 1/\mu) - k_y^2 a^2]k_0^2 + [-k_y^2 a^2 k_p^2 + k_p^4/\mu]$. This expression can be factorized as $\det(\Omega)/k_0^2 = [k_0^2 - k_p^2][k_0^2 - k_p^2/\mu + k_y^2 a^2]$. The eigenfrequencies are the same as described above.

When $k_z \neq 0$, the frequencies are given by

$$k_0^2 = [k_p^2(1 + 1/\mu) - k_y^2 a^2 + 2\,k_z^2 \pm \sqrt{R}]/2$$

where

$$R = k_y^4 a^4 + \left(2\,k_y^2 k_p^2 - 2\frac{k_y^2 k_p^2}{\mu} - 4\,k_y^2 k_z^2\right) a^2 + k_p^4 - 2\frac{k_p^4}{\mu} + \frac{k_p^4}{\mu^2}.$$

$$R = k_y^4 a^4 + k_y^2 a^2(2k_p^2(1 - 1/\mu) - 4k_z^2) + k_p^4(1 - 1/\mu)^2.$$

$$R = [k_y^2 a^2 + k_p^2(1 - 1/\mu)]^2 - 4k_y^2 a^2 k_z^2.$$

9.2.2 Medium with Two Types of Rings

We next consider the case with arrays of two types of rings, one with rings in the xz plane and the other with rings in the xy plane; the cuts of both rings are along the x axis. For the present case $\mu_{zz} = \mu_{yy} = \mu$, $\mu_{xx} = 1$, $\alpha_{yz} = -\alpha_{zy} = \alpha_0$, $\beta_{zy} = -\beta_{yz} = \beta_0$. Experimental systems can also include an array of wires along the x direction. In this section we have studied two cases: (1) the dielectric constants are diagonal but anisotropic, that along the x axis is different, $\epsilon_{xx} = \epsilon$, $\epsilon_{yy} = \epsilon_{zz} = 1$, (such as for one kind of wires); and (2) the dielectric constants are diagonal and of the same value ϵ (such as for three kinds of wires). We find that in the expression for the frequency in terms of the wave vector, α occurs in combination with the *first* power of ϵ for case (1) but with the *second* power of ϵ for case (2)! In the study of metamaterials the sign of ϵ is an important issue. This sign is irrelevant and the dispersion remains the same in case (2) We now describe our calculation in detail.

For the two-ring case the magnetoelectric coefficients become

$$\hat{\alpha} = \alpha_0 \begin{bmatrix} 0 & 0 & 0 \\ 0 & 0 & 1 \\ 0 & -1 & 0 \end{bmatrix}$$

$$\hat{\beta} = -\alpha_0 \begin{bmatrix} 0 & 0 & 0 \\ 0 & 0 & 1 \\ 0 & -1 & 0 \end{bmatrix}.$$

We thus get

$$\mathbf{k} \times \alpha = \alpha_0 \begin{bmatrix} 0 & -k_y & -k_z \\ 0 & k_x & 0 \\ 0 & 0 & k_x \end{bmatrix}$$

$$Q = \mathbf{k} \times \alpha - \beta \mathbf{k} \times = \alpha_0 \begin{bmatrix} 0 & -k_y & -k_z \\ k_y & 0 & 0 \\ k_z & 0 & 0 \end{bmatrix}$$

This operator can be written as $Q = \alpha_0 \mathbf{q} \times$ where the vector

$$\mathbf{k} = (0, -k_z, k_y).$$

For case (1), Eq. (9.3) becomes

$$(1 - \mu^{-1})(k_z \mathbf{e}_y - k_y \mathbf{e}_z)(E_z k_y - E_y k_z) + \mu^{-1}(\mathbf{kk} \cdot - k^2)$$
$$\mathbf{E} - 4\pi\omega \mathbf{k} \times (\mathbf{e}_z \alpha_{zy} E_y + \mathbf{e}_y \alpha_{yz} E_z)/c = -\omega^2 \epsilon \mathbf{E}/c^2 - 4\pi\omega[\mathbf{e}_y \beta_{yz}(\mathbf{k} \times \mathbf{E})_z$$
$$+ \mathbf{e}_z \beta_{zy}(\mathbf{k} \times \mathbf{E})_y]/c.$$

In matrix form, we get

$$\mathbf{\Omega}_l \mathbf{E} = 0 \tag{9.7}$$

where

$$\hat{\Omega}_1 = \begin{bmatrix} \mu^{-1}(k_x^2 - k^2) + \epsilon k_0^2 & \mu^{-1}k_x k_y - k_0 k_y \alpha' & \mu^{-1}k_x k_z - k_0 k_z \alpha' \\ \mu^{-1}k_y k_x + k_0 \alpha' k_y & -k_z^2 - \mu^{-1}k_x^2 + k_0^2 & k_y k_z \\ \mu^{-1}k_z k_x + \alpha' k_0 k_z & k_y k_z & -k_y^2 - \mu^{-1}k_x^2 + k_0^2 \end{bmatrix} \tag{9.7a}$$

Expanding with the help of a symbol manipulation program, we get

$$\det(\Omega_1)/k_0^2 = k_0^4\epsilon + [-\frac{k_p^2}{\mu} - \epsilon k_p^2 - 2\frac{\epsilon k_x^2}{\mu} + k_p^2 a^2]k_0^2 + [-k_p^4 a^2 + \frac{k_p^2 k_x^2}{\mu^2} + \frac{k_p^4}{\mu} - \frac{k_p^2 a^2 k_x^2}{\mu}$$
$$+ \frac{\epsilon k_x^2 k_p^2}{\mu} + \frac{\epsilon k_x^4}{\mu^2}].$$

Again $k_p^2 = k_z^2 + k_y^2$. The eigenvalues are

$$k_0^2 = [(\mu^{-1} + \epsilon - a^2)k_p^2 + 2\,\epsilon k_x^2/\mu \pm k_p^2(a^2 + \epsilon - \mu^{-1})]/2\epsilon. \tag{9.8}$$

For the plus sign, we get

$$k_0^2 = k_p^2 + k_x^2/\mu. \tag{9.8a}$$

For the minus sign, we get

$$k_0^2 = (\mu^{-1} - a^2)k_p^2/\epsilon + k_x^2/\mu. \tag{9.8b}$$

The magnetic field can be obtained from $\omega \mathbf{B}/c = \mathbf{k} \times \mathbf{E}$. We get

$$\mathbf{B} = \mathbf{k} \times \mathbf{E}/k_0 = \mathbf{k} \times \mathbf{E}/[(\mu^{-1} - a^2)k_p^2/\epsilon + k_x^2/\mu]^{1/2}.$$

As expected, the $\mathbf{B}$ field is always perpendicular to the wave vector $\mathbf{k}$. However, this is not true for the electric field.

To summarize, for case (1) with an anisotropic dielectric constant, we obtain two possible frequencies:

$$k_0^2 = k_z^2 + k_y^2 + k_x^2\mu^{-1}, \quad k_0^2 = v^2(k_z^2 + k_y^2) + k_x^2/\mu \tag{9.9}$$

where

$$v^2 = v_0^2 - \alpha'^2/\epsilon. \tag{9.9a}$$

$k_0 = \omega/c$, $\alpha' = 4\pi\alpha$, $v_0^2 = 1/(\mu\epsilon)$ is the group velocity when the magneto-electric effect is absent. For the second mode, the velocity v in Eq. (9.9a) now contains a term $-\alpha'^2/\epsilon$. As we shall see below, for the case with three types of rings, the dispersion relation also contains a term proportional to α'^2 but in that case it is the *square* of ϵ that appears. Since α possesses a significant imaginary part, if ϵ is positive, this term can increase the phase speed of light and make the speed real even if μ is negative. If ϵ is negative, this term will now lower the speed of light, opposite to case with three types of rings! As we shall see, this opposite trend comes from the anisotropic nature of the dielectric constant. We next look at the polarization of these modes.

For the first mode, the electric field is along $\mathbf{e}_q$ that is perpendicular to both $\mathbf{k}$ and $\mathbf{e}_x$.

$$\mathbf{E} = E_q\mathbf{e}_q \tag{9.9b}$$

For the second mode, the electric field is perpendicular to $\mathbf{e}_q = (0, -k_z, k_y)/(k_z^2 + k_y^2)^{1/2}$:

$$\mathbf{E} = E_k\mathbf{e}_k + E_p\mathbf{e}_p \tag{9.9c}$$

where $\mathbf{e}_{p=}\mathbf{e}_k \times \mathbf{e}_q, E_p/E_k = [k_x^2(\epsilon - 1)/k^2 + 1]k_0/[q\,\alpha' - k_0(\epsilon - 1)p_xk_x/(kp)]$.

The electric field now has a *longitudinal* component along the direction of the wave vector. Because α is mostly imaginary the longitudinal and the transverse components of the electric field are now out of phase: the electric field is elliptically polarized.

9.2.2.1 Isotropic ϵ

We next address the case with an isotropic dielectric constant tensor. We obtain the eigenvalue equation

$$\mathbf{\Omega E} = 0$$

where

$$\hat{\Omega} = \begin{bmatrix} \frac{-k_y^2-k_z^2}{\mu} + \epsilon k_0^2 & \frac{k_xk_y}{\mu} - k_0k_y\alpha' & \frac{k_xk_z}{\mu} - k_0k_z\alpha' \\[2ex] +k_0k_y\alpha' & -\frac{k_x^2}{\mu} - k_z^2 + \epsilon k_0^2 & k_yk_z \\[2ex] +k_0k_z\alpha' & k_yk_z & -\frac{k_x^2}{\mu} - k_y^2 + \epsilon k_0^2 \end{bmatrix}$$

From this we obtain the quadratic equation

$$\det(\hat{\Omega})/k_0^2 = k_0^4\epsilon^3 + k_0^2\epsilon[k_p^2(\alpha'^2 - \epsilon - \frac{\epsilon}{\mu}) - 2\,\frac{\epsilon k_x^2}{\mu}] + k_p^2(\frac{\epsilon k_x^2}{\mu} - \frac{\alpha'^2 k_x^2}{\mu} + \frac{\epsilon k_x^2}{\mu^2}) + \frac{k_p^4\epsilon}{\mu}$$
$$+ \frac{\epsilon k_x^4}{\mu^2} - k_p^4\alpha'^2$$

The frequencies for the two modes are given by

$$k_0^2 = (k_z^2 + k_y^2 + k_x^2\mu^{-1})/\epsilon, \ k_0^2 = v^2(k_z^2 + k_y^2) + k_x^2/(\epsilon\mu) \tag{9.10}$$

where now

$$v^2 = v_0^2 - \alpha'^2/\epsilon^2. \tag{9.10a}$$

It is the square of the dielectric constant that appears in Eq. (9.10a), different from (9.9a). Thus for this velocity, the sign of the dielectric constant is immaterial.

9.2.3 3-Ring Medium

We next discuss the case with three types of rings with cuts along three different axis. The configuration we have in mind is illustrated in Fig. 9.1. Each set of rings can have different "polarities", depending on whether the cut of the ring is along the + or the − direction. We have considered systems with different polarities and found the physics to be similar. Here we consider the case when the polarities of the three kinds of rings are the same. This system is non-conventional because the magnetoelectric coefficients $\hat{\alpha}$, $\hat{\beta}$ are off-diagonal tensors. Their diagonal components and some of the off-diagonal components are zero. Systems similar to this have been studied experimentally by the Boeing group where additional arrays of wires along three orthogonal directions are also present. Our calculation can be applied to such situations.

For a single split ring placed on the xy plane with a cut opened at $\varphi = 0$, the results of Chaps. 3 and 4 show that an electric field E_y generates a magnetic moment m_z, while a magnetic field H_z generates an electric dipole moment p_y. For the ring structure illustrated in Fig. 9.1, it is straightforward to find the magnetoelectric tensor as

$$\hat{\alpha} = \alpha_0 \begin{bmatrix} 0 & 0 & 1 \\ 1 & 0 & 0 \\ 0 & 1 & 0 \end{bmatrix} \,,\quad \hat{\beta} = -\alpha_0 \begin{bmatrix} 0 & 1 & 0 \\ 0 & 0 & 1 \\ 1 & 0 & 0 \end{bmatrix}.$$

For example, for a ring in the yz plane with a gap along the y direction, an electric field E_z will produce a magnetic moment m_x. This corresponds to the first row of the $\hat{\alpha}$ matrix. In the same way, the second and third rows of this matrix come from rings in the xz, xy planes with gaps along the z and x axis, respectively.

It is mathematically convenient to express the cross product as a matrix multiplication. After some straightforward algebra, we find that

$$\mathbf{Q} = \mathbf{k} \times \hat{\alpha} - \hat{\beta}\mathbf{k}\times = \alpha_0 \begin{bmatrix} 0 & k_y & -k_x \\ -k_y & 0 & k_z \\ k_x & -k_z & 0 \end{bmatrix} = \alpha_0 \mathbf{q}\times$$

where the vector $\mathbf{q} = -(k_z, k_x, k_y)$. Eq. (9.3) thus becomes

$$\mu^{-1}[\mathbf{k}\mathbf{k}\cdot -k^2]\mathbf{E} - 4\pi\omega\alpha\mathbf{q} \times \mathbf{E}/c = -\omega^2\epsilon\mathbf{E}/c^2 \tag{9.11}$$

Ordinarily the electric field lies in the plane perpendicular to $\mathbf{k}$. This is not necessarily true anymore. To make a connection with the conventional result, we find it convenient to represent the components of the electric field in terms of the following three mutually perpendicular vectors. Define $\mathbf{p} = \mathbf{q} \times \mathbf{k}$ that is perpendicular to both $\mathbf{q}$ and $\mathbf{k}$. Then construct the component of k that is perpendicular to q: $\mathbf{k}' = \mathbf{k} - \mathbf{q}(q \cdot k)/q^2$.

$$\mathbf{E} = E_k\mathbf{e_{k'}} + E_q\mathbf{e_q} + E_p\mathbf{e}_p;$$

Equation (9.3) becomes

$$\hat{\Omega}\mathbf{E} = 0 \tag{9.12}$$

where

$$\hat{\Omega} = \begin{bmatrix} k'^2 + x & k_q k' & \gamma k_0 k \\ k_q k' & k_q^2 + x & 0 \\ -k_0 k \gamma & 0 & +x \end{bmatrix}, \tag{9.12a}$$

$x = k_0^2 - k^2$, $k_q = \mathbf{e_q} \cdot \mathbf{k}$, $k_0 = (\mu\epsilon)^{1/2}\omega/c$, $\gamma = 4\pi\alpha\sqrt{\mu/\epsilon}$. After some algebra, we find that

$$\det(\Omega) = k_0^2[x^2 + (\gamma k)^2(k_q^2 + x)].$$

The condition $\det(\Omega) = 0$ leads to a quadratic equation in x, we obtain the dispersion

$$k_0^2 = k^2 - 0.5\left((\gamma k)^2 \pm [(\gamma k)^4 - 4(\gamma k k_q)^2]^{1/2}\right). \tag{9.13}$$

We next explore the implication of this result.

The physics is particularly simple for $k_q = 0$, or close to a resonance when γ becomes large and the term proportional to k_q can be neglected. In that case from Eq. (9.12) the electric field along $\mathbf{q}$ is not coupled to the other two components. The condition $k_q = 0$ is obtained, for example, if k is along one of the axis, e.g., $k_x = k_y = 0$, the directions $\mathbf{k'}$, $\mathbf{q}$, $\mathbf{p}$ then corresponds to the directions $\mathbf{z}, -\mathbf{x}, \mathbf{y}$, respectively. In this limit, the dispersion for the two normal modes are given by

$$k_0^2 = k^2 v_0^2, \quad k_0^2 = k^2 v^2. \tag{9.14}$$

$$v^2 = v_0^2 - \alpha'^2/\varepsilon^2. \tag{9.14a}$$

We call these two modes the ordinary mode and the extraordinary mode. The ordinary mode corresponds to a $-$ sign in Eq. (9.13) and is one that we normally expect. It is transverse, the electric field is polarized along q. This is illustrated by the field E_- in Fig. 9.1. The second mode in Eq. (9.14) behaves quite anomalously. The velocity in Eq. (9.14a) is the same as that in Eq. (9.10a). Because α possesses a significant imaginary part (see Chap. 4), the factor $-(4\pi\alpha)^2/\epsilon^2$ provides for an increase in the phase velocity if ϵ is real. Even when $\mu\epsilon$ is negative, the second factor can render the real part of ω^2 positive close to the resonance. This can be seen in Fig. 9.2 where we have evaluated $(v/c)^2$ using the value of α obtained in Chaps. 3 and 4. On both sides of the resonance where μ changes sign, the real part of v^2 remain positive. Very close to the resonance, the real part of α becomes large. Although the product $\mu\epsilon$ is positive, $\text{Re}[v^2]$ can be less than zero, as is shown in Fig. 9.2.

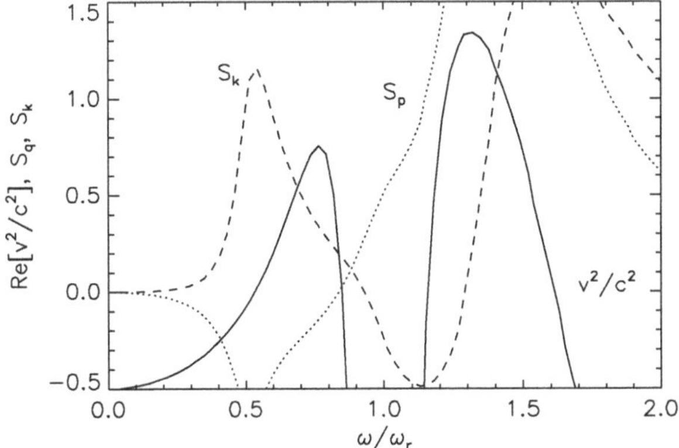

Fig. 9.2 Illustrative result for the transverse and longitudinal Poynting vectors S_p, (*dotted line*) S_k (*dashed line*) normalized by $c|E_p|^2/(4\,\pi)$ and the square of the velocity (*solid line*) as a function of the frequency normalized by the resonance frequency. $R^3/V = 0.03$, $\epsilon = -2$, the conductor resistance $2\pi R r_c = 0.1 Z_0$ where $Z_0 = (\mu_0\,\varepsilon_0)^{1/2} = 377\Omega$ is the impedance of the vacuum (from Ref. [8])

For the extraordinary mode (2nd mode in Eq. (9.14)), we get from Eq. (9.12) that

$$\mathbf{E} = E_k \mathbf{e}_k + E_p \mathbf{e}_p; \tag{9.15}$$

$E_k/E_p = -\gamma k/k_0$. Because γ possess a significant imaginary part the two components are $90°$ out of phase; the electric field is *longitudinally* elliptically polarized in the plane formed by $\mathbf{p}$ and $\mathbf{k}$! This is illustrated by E_+ in Fig. 9.1. The corresponding magnetic field is given by $\mathbf{B} = \mathbf{k} \times \mathbf{E} c/\omega = ckE_p \mathbf{e}_q/\omega$ and is transverse.

In general for a nonzero k_q, from Eq. (9.13), we get $E_q = -E_{k'} k_q k'/(k_q^2 + k_0^2 - k^2)$, $E_p = -E_{k'} k_0 k\gamma/(k_0^2 - k^2)$. The electric field has a component in the k–q plane and another component that is $90°$ out of phase in a direction perpendicular to the k–q plane. Thus it is also longitudinal elliptically polarized.

In previous studies a key question is the direction of energy flow, which is given by the Poynting vector $S = \text{Re}[c\mathbf{E} \times \mathbf{H}^*/(4\pi)]$. Kamenetskii [9] has shown that for bianisotropic media, Poynting's theorem has the continuity equation form and energy transport is possible if the envelop function of the wave packets satisfy certain reasonable conditions. Because of the longitudinal elliptic polarization, both $\mathbf{E}$ and $\mathbf{H}$ now have components $E_\|$, $H_\|$; $E_\perp$, $H_\perp$ along and perpendicular to $\mathbf{k}$, the cross product of $\mathbf{E}$ and $\mathbf{H}$ can have contributions $\text{Re}[E_\| H_\perp*]$, $\text{Re}[E_\perp H_\|*]$ perpendicular to $\mathbf{k}$. When damping is included, $E_\|$ $(E_\perp)$ and $H_\perp$ $(H_\|)$ contains contributions that are out of phase with each other and thus give rise to a nonzero component $\mathbf{S}_p$ of the Poynting vector that is perpendicular to the wave vector. More precisely, substituting in the expressions for the EM fields, we obtain

$$\mathbf{S} = \mathrm{Re}[\mathbf{S}_0 + \mathbf{S}']$$

where $\mathbf{S}_0 = c\mathbf{E} \times [\boldsymbol{\mu}^{-1}\mathbf{B}]^*/(4\pi)$, $\mathbf{S}' = c\mathbf{E} \times [\boldsymbol{\alpha}\mathbf{E}]^*/(4\pi)$. There is now an additional term. For the particular example with the wave vector along the z axis discussed above

$$(4\pi/c)\mathbf{S}'/|E_p|^2 = \alpha^*(\mathbf{e}_q + |\gamma^2/(1-\gamma^2)|\mathbf{e}_p + \mathbf{e}_k[\gamma/(1-\gamma^2)^{0.5}]^*).$$

$$(4\pi/c)\mathbf{S}_0/|E_p|^2 = [(\epsilon/\mu)^{0.5}/(1-\gamma^2)^{0.5}]^*[\gamma/(1-\gamma^2)^{1/2}\mathbf{e}_p + \mathbf{e}_k)].$$

Both $\mathbf{S}_0$ and $\mathbf{S}'$ now has a component *perpendicular* to $\mathbf{k}$. This component becomes significant close to the resonance when the real part of α becomes large, as is illustrated in Fig. 9.2 where we show the transverse and the longitudinal components S_p, S_k.

Photonic crystals and anisotropic materials exhibit negative refraction because the group velocity $\mathbf{v}_g = \nabla \cdot \omega$ is not parallel to the wave vector. The perpendicular direction of energy flow discussed here does not come from the group velocity, which is along the wave vector for the example discussed here. Indeed, differentiating Eq. (9.13), we get $2k_0\mathbf{v}_g/c = 2k\mathbf{e}_k - \gamma^2 k\mathbf{e}_k + (-)[\gamma^4 k^3 \mathbf{e}_k - 2\gamma^2\left(kk_q^2\mathbf{e}_k + k^2 k_q \mathbf{e}'\right)]]/[(\gamma k)^4 - 4(\gamma kk_q)^2]^{1/2}]$ where $\mathbf{e}' = -(k_z + k_y, k_x + k_z, k_x + k_y)/k$. For $k_q = 0$; the coefficient of the $\mathbf{e}'$ term is zero, the group velocity is along the wave vector.

Our effect is present only in the presence of damping, when there is no exact theorem that requires the Poynting vector to be parallel to the group velocity. Usually we expect a smearing in the direction of energy flow in the presence of damping. Here we find the presence of additional terms in the perpendicular direction.

9.3　Reflection and Refraction from Composites of Split Rings

The new kind of longitudinal elliptic polarization has other unexpected consequence. The ordinary mode and the extraordinary mode exemplified in Eqs. (9.13) and (9.14) can be coupled at an interface. To illustrate the physical phenomenon, we consider the case of composites with two types of rings. An exact solution in the homogenized continuous model medium characterized by the effective susceptibilities for a particular two-ring medium of case (1) was performed. When a TM-polarized 2D Gaussian beam is incident from air on its top surface, it is found that the refracted wave inside the medium splits into two beams, as is depicted in Fig. 9.3a. One of the beams corresponds to the ordinary mode with the $\mathbf{E}$ field perpendicular to the $\mathbf{k}$ vector, as is schematically illustrated in Fig. 9.3c. The other beam corresponds to the extraordinary mode with the $\mathbf{E}$ field rotating in the $\vec{x} - \vec{k}$ plane, as is schematically shown in Fig. 9.3b.

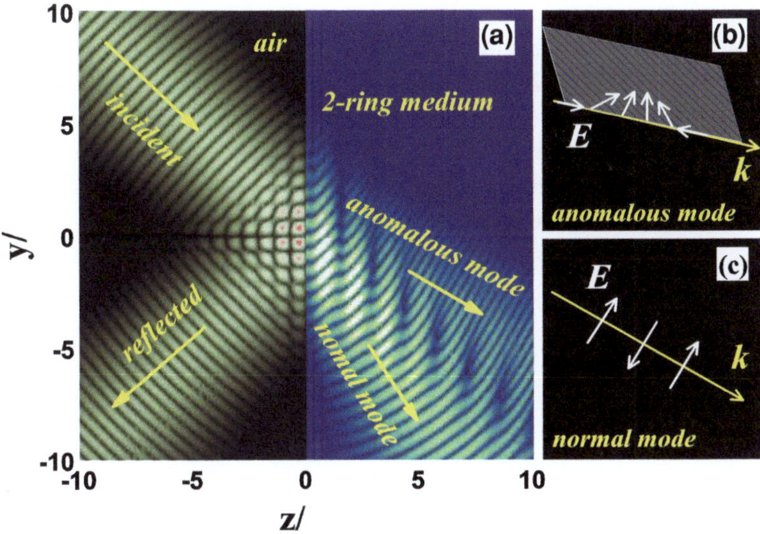

Fig. 9.3 a Illustration of wave reflection/refraction as a 2D 2λ-wide Gaussian beam with a TE polarization ($\mathbf{E}\|\mathbf{x}$) strikes on a two-ring medium with $\varepsilon_{xx} = 10$, $\varepsilon_{yy}\ \varepsilon_{zz} = 2$, $\mu_{xx} = 1$, $\mu_{yy} = \mu_{zz} = 1.1$, $\alpha_0 = 5i/(4\pi Z_0)$. **b** Expanded view of the anomalous beam with $\mathbf{E}$ vector rotating on the $\mathbf{x}$–$\mathbf{k}$ plane. **c** Expanded view of the normal beam with $\mathbf{E}$ perpendicular to $\mathbf{k}$ (from Ref. [8])

For a TE-polarized Gaussian beam, no such double refraction is found. This can be understood from the boundary conditions that the tangential components of $\mathbf{E}$ and $\mathbf{H}$ are continuous. Because the anomalous refracted beam is longitudinal elliptically polarized with a $\mathbf{E}$ component lying on the y–z plane, a TM-polarized incident wave can excite not only the normal TM mode, but can also the extraordinary mode which possesses a $\mathbf{E}$ component parallel to the interface. In contrast, a TE-polarized incident wave can only excite this anomalous mode since its $\mathbf{E}$ vector is strictly perpendicular to that of the normal refracted beam. To further illustrate the physics, we next provide the detail algebra of the simple matching of the boundary fields in the long wavelength limit, again for the composite with two types of rings.

Consider an example of incoming light with the wave vector in the zy plane, i.e. $\mathbf{k}_i = (0, k_y, k_{iz})$. Now k_y is conserved. For the incoming wave, consider the case where the electric field is along x: $E_i = (E, 0, 0)$; $H_i = B_i = E(0, k_{iz}, -k_y)/k_i$. For the reflected with, $k_l = (0, k_y, -k_{iz})$. The reflected electric field is perpendicular to this wave vector, the most general form can be written as $E_l = (f, dk_{iz}/k_i, dk_y/k_i)$ with constants f and d to be determined. The corresponding reflected magnetic field is given by Maxwell's equations as $H_l = B_l = (k_y dk_y/k_i + k_z dk_z/k_i, -k_{iz}f, -k_y f)/k_i = (d, -k_{iz}f/k_i, -k_y f/k_i)$. The polarizations of the two normal modes inside the split ring composite are given in Eq. (9.9b) and (9.9c). Now $\mathbf{e}_\mathrm{p}$ is along the x direction. For the transmitted beam, we get a linear combination of the two polarizations $\mathbf{E}_t = a(E_x\mathbf{e}_x + E_p\mathbf{e}_k) + b\mathbf{e}_q = (aE_x, (aE_pk_y + bk_z)/k, (aE_pk_z - bk_y)/k)$ with constants a and b to be determined. The corresponding magnetic field is given by $\mathbf{B} = [k_y(aE_pk_z - bk_y)/k - k_z(aE_pk_y + bk_z)/k, k_zaE_x/k, -k_yaE_x]/k_i$. This can be simplified as $\mathbf{B} = [-bk, k_zaE_x/k, -k_yaE_x]/k_i$; $\mathbf{H} = [-bk, \mu^{-1}k_zaE_x/k, -mu^{-1}k_yaE_x]/k_i$. The tangential components are in the xy plane. The projections are: For the transmitted beam: $\mathbf{E}_t = (aE_x, (aE_pk_y + bk_z)/k, 0)$. $\mathbf{H} = [-bk, \mu^{-1}k_zaE_x/k, 0]/k_i$. For the reflected beam: $E_l = (f, dk_{iz}/k_i, 0)$, $H_l = (d, -k_{iz}f/k_i, 0)$. For the incoming beam: $E_i = (1, 0, 0)$; $H_i = (0, k_{iz}, 0)/k_i$.

Equating the tangential components, we get the equations:

$$1 + f = aE_x,$$

$$dk_{iz}/k_i = (aE_pk_y + bk_z)/k,$$

$$d = -kb/k_i,$$

$$k_{iz}/k_i - k_{iz}f/k_i = \mu^{-1}k_zaE_x/k.$$

There are four equations for four unknowns. We get

$$a = 2/(E_x[1 + \mu^{-1}k_zk_i/(kk_{iz})]),$$

$$f = 2/[1 + \mu^{-1}k_zk_i/(kk_{iz})] - 1,$$

$$b = -a(E_pk_y)/(k_{iz}k^2/k_i^2 + k_z).$$

$$d = -kb/k_i. \tag{9.16}$$

As is advertized, both a and b are nonzero!

In conclusion, we find that for a collection of split rings, because of its anisotropic off-diagonal magnetoelectric tensor, many interesting new phenomena remain to be discovered, with or without additional arrays of wires. Examples of these are (1) onset of transmission even for a negative μ, or no transmission even when both ϵ and μ are negative, (2) perpendicular transport of light, (3) anomalous behaviors in refraction and reflection.

9.4 Propagation of Electromagnetic Waves in Helixes

So far we have focused on non-chiral systems with only off-diagonal magneto-electric coefficients. Here we consider an example of a chiral system with finite diagonal magnetoelectric coefficients. The properties of a single helix and their magnetoelectric coefficients are discussed in Chap. 4. Here we explore the physics of light propagating through a composite of helixes all oriented parallel to the z axis. To illustrate the essential physics, we make the simplest approximation so that the susceptibilities are approximately the average value of that of the matrix and the helixes weighted by the corresponding volume fractions. An improved estimate can be obtained with the Claussius-Mosotti approximation or variations thereof.

To illustrate our effect, we first consider the case so that the helix has an extra half turn, and the azimuthal angle is restricted to the range $0 < \varphi < 3\pi..$ As is discussed in Chap. 4, in that case only α_{zz} is nonzero. For plane waves, Eq. (9.3) becomes

$$\mathbf{k} \times \mu^{-1}\mathbf{k} \times \mathbf{E} - 4\pi\omega\mathbf{k} \times \mathbf{e}_z\alpha_{zz}E_z/c = -\omega^2\epsilon\mathbf{E}/c^2 - 4\pi\omega\beta_{zz}\mathbf{e}_z(k \times E)_z/c. \quad (9.17)$$

We look for a solution with $\mathbf{k}$ along x and $\mathbf{E} = E_z\mathbf{e}_z + E_y\mathbf{e}_y$. We get in matrix form, $\mathbf{\Omega E} = 0$ where ($k_0 = \omega/c$, $\alpha' = 4\pi\alpha$, $\beta' = 4\pi\beta$)

$$\hat{\Omega} = \begin{bmatrix} -\mu^{-1}k^2 + \epsilon_{yy}k_0^2 & k_0k\alpha' \\ -k_0k\alpha' & -k^2 + \epsilon_{zz}k_0^2 \end{bmatrix}. \quad (9.18)$$

The condition

$$\det(\Omega) = 0 \quad (9.18a)$$

leads to the equation

$$2\mu^{-1}(v_\pm/c)^2 = b \pm [b^2 - 4\mu^{-1}\epsilon_{yy}\epsilon_{zz}]^{1/2}, \quad (9.19)$$

$$E_z/E_y = -k_0k\alpha'/[-k^2 + \epsilon_{zz}k_0^2].$$

where $v = \omega/k$, $b = (\mu^{-1}\epsilon_{zz} + \epsilon_{yy} - \alpha'^2)$. α' is imaginary. Hence E_z, E_y are $90°$ out of phase; the light become elliptically polarized in the transverse direction. The speed of light $v_\pm$ is significantly different for the two polarizations $\pm$. This difference in velocity leads to the giant Faraday effect discussed in Chap. 4.

References

1. L.D. Landau, J.L. Lifshitz, *Electrodynamics of Continuous Media* (Pergamon, Oxford, 1960), p. 119
2. I.E. Dzyaloshsinski, Sov. Phys. JETP **10**, 628 (1960)
3. D.N. Astrov, Sov. Phys. JETP **11**, 708 (1960)
4. J.G. Wan, J. M. Liu, G.H. Wang, C.W. Nan, Appl. Phys. Lett. **88**, 182502 (2006)
5. L. Jiang et al., Science **302**, 661 (2004)
6. I.V. Lindell, A.H. Sihvola, S.A. Tretyakov, A.J. Vitanan, *Electromagnetic Waves in Chiral and Bi-isotropic Media* (Artech House, Boston, 1994)
7. J. Pendry, Science **306**, 1353 (2004)
8. S.T. Chui, W.H. Wang, L. Zhou et al., Longitudinal elliptically polarized electromagnetic waves in off-diagonal magnetoelectric split-ring composites. J. Phys. Condens. Matter **21**(29), article number 292202 (2009)
9. E.O. Kamenetskii, Phys. Rev. **E54**, 4359 (1996)

Index

A

Absorber, 57, 63, 99
Anisotropic, 42, 52, 122, 125, 127, 129,
    134, 136
    see also Isotropic
Antenna, 7, 45, 54–57, 63, 101
    dipole, 9, 45, 53, 101
    Yagi, 63
"Antibonding" mode, 85
    see also "Bonding" mode

B

Bi(-)anisotropic, 8, 14, 21, 29, 32, 34–36, 42,
    121, 122, 133
    see also Magnetoelectric
"Bonding" mode, 85
Boundary condition, 13–16, 64, 66
Breakdown field, dielectric, 99, 100, 104
Broadside component/radiation/mode, 54, 57
    see also End(-)fire component/radiation/
    mode

C

Capacitance (matrix), 2, 11, 19, 39, 45, 49, 52,
    64–66, 70, 103
    diagonal, 30
    mutual, 2, 30–32, 39, 40, 67, 76, 85, 96, 102
    self, 2, 30, 39, 40, 99, 102
    symmetrical, 30
    see also Capacitive (electric) field,
    Capacitive term

Capacitive (electric) field, 3, 4, 10, 11,
    12, 46, 65
Capacitive term, 13, 49
Chiral, 46, 55, 58, 59, 110, 123, 136
Circuit impedance, 2, 3, 16
    see also Capacitance (matrix), Inductance
    (matrix), Impedance/circuit parameters
    matrix
Circular polarizer, efficient, 45
Claussius-Mosotti approximation, 59, 137
Current distributions, 39

D

Damping, 13, 26, 57, 65, 106, 108, 133, 134
Dark mode, 80, 87
Dielectric constants, diagonal, 127

E

Eigenmode, 1–3, 16, 21, 60, 64, 68, 69, 72–77,
    79–84
    $2m$th resonance, 20
    odd-numbered, 20
    see also Eigenstate
Eigenstate, 18
Eigenvector
    (anti-)symmetrical, 20
End(-)fire component/radiation/mode, 54, 57
End/boundary (electric) field, 15, 67–71,
    77–79, 81, 91, 96–98, 100, 103
    See also Local/localized (end/boundary)
    (electric) field

S. T. Chui and L. Zhou, *Electromagnetic Behaviour of Metallic Wire Structures*,
DOI: 10.1007/978-1-4471-4159-4, © Springer-Verlag London 2013